Thinking Outside the Brain Box

Thinking Outside the Brain Box

Why Humans are not Biological Computers

ARIE BOS

Floris Books

Translated from Dutch by Philip Mees

First published in Dutch as *Mijn brein denkt niet, ik wel* by Uitgeverij Christofoor, Zeist, Netherlands in 2015
First published in English by Floris Books in 2017

© 2015 Uitgeverij Christofoor, Zeist
English version © 2017 Floris Books

All rights reserved. No part of this publication may be reproduced without the prior permission of Floris Books, Edinburgh
www.florisbooks.co.uk

Also available as an eBook

British Library CIP data available
ISBN 978-178250-428-3
Printed by TJ International

Contents

Foreword 6

Part 1: My Brain Doesn't Think
1. How Come the Brain is so Smart? 12
2. The Brain that Rewires Itself 24
3. What are Mirror Neurons? 29
4. The Instrument of Consciousness 40
5. How the Pictures are Interpreted 50
6. Where Does Consciousness Reside? 61
7. The Brain in a Vat 83
8. Consciousness Needs the Body 89
9. The Brain and Human Freedom 101
10. No Free Will in the Lab 115
11. The Human Being as a Machine 137
12. Two Hemispheres Under One Roof 151
13. Conflict Between Neighbours? 162
14. The Role of the Brain in Thinking 174

Part 2: I'm The One Who Thinks
15. To Be or Not To Be 182
16. Do 'I' Exist? 202
17. Afterword: Is there an Alternative to Materialism? 219

Notes 235

Foreword

'I find it a very reassuring idea that we are our brain,' said the lady.
'How so?' I asked.
'Well, I'm really addicted to cigarettes, and now I know that it's simply something genetic in my brain.'
'And you find that reassuring?'
'Yes, my brain simply longs for nicotine; there's nothing I can do about it.'
'So your brain is actually very happy to be addicted?'
'I guess you could say that.'
We drank. An important problem seemed to have been solved.
'And you? Are you also happy to be addicted?'
'No, of course not. It's extremely bad for your health, but what can you do about it?'

That was a snippet of a conversation at a party, a real contemporary conversation, because the brain is a hot topic. The lady was quoting from the title of Dick Swaab's wildly successful book, *We Are Our Brain*,[1] but she had ostensibly missed the drift of it. For of course you can't be your brain and at the same time want something other than the brain wants. And, according to the title of the book, self and brain coincide. Evidently, it is difficult to truly recognise the consequence of that. As we will see in this book, this not only holds for laymen, but also for some neuroscientists and neurophilosophers who, each in their own way, proclaim that we are our brain.

Actually, the lady at the party precisely evoked the problem this book addresses: are we capable of separating ourselves from our brain, doing something different from what the brain 'wants?' In this book I try to make a credible case that this is indeed the case.

Otherwise, what purpose would our self-reflecting consciousness serve? Before making the attempt, however, I first have to make a scientific-philosophical statement: except in mathematics, science does not *prove*, it *makes credible*. It shows connections, correlations. What is much more difficult is to demonstrate or prove what is cause and effect in these correlations. Causality is always an interpretation.[2]

The Fly

Certainly there are lots of creatures that are their brain – the fly, for instance, that is flying round and round in my room right now, and hits its head against the window all the time. It is doing what flies have done for ten thousand or perhaps a hundred thousand years: nothing but what their tiny little brains tell them to do. And those tiny little brains provide for a behaviour that has allowed them to survive in their eco-system. Only at that time there were no windows, and the little creatures haven't learned much since then.

But we are born with brains that still have to learn an awful lot of things, and are formed by such learning. We are therefore not our brain, but produce it ourselves, so to speak. From the beginning of evolution, the brain has been principally occupied with the body. This holds not only for flies, but also for us and some forms of bodily behaviour such as eating, coughing, throwing up, relieving ourselves, and so on. But even these are things human beings have to learn to control. Behaviour and the considerations connected with it were the next step in the evolution of the brain, and in human beings this has come to play an incomparably large role. Children are quickly able to do very different things than their parents, while their brains are much the same.

Neurodeterminism

It appears that, apart from a number of dissidents, most researchers are convinced that the brain produces consciousness; and that is also true for most other people in the western world. I don't believe that this view has an official name in neurophilosophy. In this book

I will call it 'neurodeterminism,' meaning: we are wholly determined by our brain. For many people this is evidently a huge relief for it takes care of any inhibitions in accepting themselves. You simply are who you are, and there is nothing you can do about it; your brain is responsible. Witness the lady at the party, with which this foreword began, for whom all reason to be ashamed of herself had fallen away. It also makes it unnecessary to look for something 'higher' or 'deeper,' or 'self': ultimately it is all about chemical and electrical processes in the brain. In brief, if we are our brain, we can feel liberated.

Not everyone, however, has this sense of relief, and many people feel uncomfortable with the concept. For, as will be extensively demonstrated in the course of this book, the view that we are our brain gives rise to a whole flood of related statements in our view of life such as 'there is no free will,' 'we are automatons,' 'consciousness plays no role in what we do,' 'consciousness does not give a dependable picture of reality,' 'consciousness talks rubbish,' 'altruism is just veiled selfishness,' 'the "I" is an illusion.' And it all has to do with the fact that in neurodeterminism we are no longer beings, but things: brains.

The Theme of This Book

But if we are not our brain, then what are we? Our body? Our consciousness? Both of these? Or do we *have* a body and consciousness? If so, who or what is the owner? And what is then the role of the brain in the existence of consciousness? It is not easy to arrive at a consensus on these questions. And that is exactly what this book is about, about the question whether we can be 'boss in our own brain,' and are therefore indeed responsible in some way or other for what goes on there; and thus also about free will. In addition, it is about the questions of why, in spite of the objections raised above, it appears so logical to think that our brain arranges everything; why we have come to understand the world and ourselves so mechanistically; and why some neuroscientists insist that we are but automatons. These are the questions considered in *Part 1: My Brain Doesn't Think*.

But most of all, the book is about the question: what are we really? In language we have no trouble naming what we are. If you ask me what I am, I say: 'I; I am I.' Saying that, I mean exactly the same thing as anyone else who calls himself 'I.' We cannot communicate without personal pronouns: I, you, she, we, etc. Ordinarily there is no need to ask what we mean by that; we certainly don't mean something like 'my brain wants to go on vacation in Spain this year.' Only in philosophy and neuroscience is this 'I' made into a problem. And because in these disciplines it is also necessary to communicate, it gets a different name: the *self*, so that the author can simply continue to use the pronoun 'I' when he refers to himself. I will follow this practice.

Part 2: I'm The One Who Thinks, is about the question of whether this self really exists. In the last chapter I consider whether there is an alternative to the mechanistic view of the world.

This book is principally about the relationship between the brain and consciousness. Perhaps I should make clear right here what I mean with the word consciousness, to satisfy those who demand a definition before talking about a concept. Actually, this is strange. No one demands a definition of 'brain' although very few people have actually had a brain in their hands. But consciousness, which we all know better than anything else in the world – for after all, we know everything we know exclusively thanks to this consciousness – is hard to circumscribe. In this book I use the word consciousness for everything we experience consciously, and also unconsciously – our perceiving, thinking, feeling and acting. For me, all we experience unconsciously is also part of consciousness. This implies that animals also possess consciousness. But not automatons, they don't experience anything. For that matter English has a practical word: the mind. One can speak about *the conscious mind* and *the unconscious mind*. Much less confusing than *conscious consciousness* and *unconscious consciousness*.

There are of course people who think that all of this is merely about ideas, which are at any rate unimportant for daily practical life. But ideas are not 'merely opinions.' In Chapter 10, I will describe research that shows that ideas very distinctly influence our behaviour. But we could have known that before. Just think of countries that are based on ideas of how people are, or should be – ideas with which their subjects are indoctrinated, irrespective of whether they are communist, fascist, capitalist, socialist, Moslem, Christian or Hindu states. The ideas determine to a large extent how people relate to each other and what values they cultivate. In this book I want to turn away from one such idea, namely that only matter determines reality. I hope to make credible that this is but a prejudice that influences hypotheses about cause and effect in a one-sided way. The advantage this gives me is that I don't need to use the word illusion when I want to explain something that does not admit of a materialistic explanation. It therefore allows me a much broader and, in my opinion, more realistic view of the world and the human being, as will become clear in Chapter 17.

Part 1

My Brain Doesn't Think

CHAPTER 1

How Come the Brain is so Smart?

We may justifiably say that what we do depends on who we are, but we have to add that, to a certain extent, we are what we do, and we constantly create ourselves.

Henri Bergson, *L'évolution créatrice*

Some years ago, a patient with a rather intimidating musculature left the office of a colleague who used a room in our office to coach difficult cases under the government's social legislation. He straightaway rang his employer to tell him the doctor had just confirmed he was ill. Evidently, his boss was not happy with that, and everyone in the waiting room witnessed how quickly a labour problem can escalate. The man got so mad that with a karate kick he destroyed a beveled nineteenth century windowpane in the hallway. When I ran up to him with the cautious words: 'What are you doing?' he spoke the unforgettable words: 'There's nothing I can do about it. I didn't make myself, did I?'

Of course, he did not mean I should present his family with the bill for the damage but he wanted to emphasise that he was not responsible for what he did. This was just the way he was. At first sight he seemed to be living proof of the thesis defended by psychiatrist Theodore Dalrymple, that some opinions arising in academic circles quickly find their way into all levels of society. For this attitude closely resembles what the overwhelming quantity of 'brain books' proclaim that are currently on the market.[1]

But this event happened in 1990 when those books had not yet appeared. One of the few authors who had then already written something of this sort was the successful scientist Francis Crick who in 1953, together with James Watson, discovered the structure of DNA, which they then called 'the secret of life.' At the time, this seemed to be the perfect answer to all questions but, looking back, it was a little premature, for a cell of which the nucleus has been removed can continue to live for weeks.

The Astonishing Hypothesis

After he thought he had solved the riddle of life, Crick threw himself into the other great riddle: the neurosciences. In 1994 he published *The Astonishing Hypothesis*, in which he summarised this astonishing hypothesis in one sentence:

> You, your joys and your sorrows, your memories and ambitions, your sense of personal identity and free will, are in fact no more than the behaviour of a vast assembly of nerve cells and their associated molecules ... This hypothesis is so alien to the ideas of most people today that it can truly be called astonishing.[2]

I may of course be mistaken, but I don't think our window kicker had read this book. To him, Crick's hypothesis might not have been so astonishing. What's more, for most people it wouldn't either. On the basis of an improvised, non-representative poll of fellow physicians and friends, I have reason to think that Crick's view reflects long-established public opinion. It is at any rate a very simple model to understand. It boils down to the idea that our brain produces our consciousness (Crick's joys, delights, laughter and sports, and sorrows, grief, despondency, and lamentations), just as radios produce music, television produces films, and computers produce knowledge (if indeed they actually do that).

Otherwise, where would our consciousness come from? And also, when part of the brain is damaged, a person loses part of his cognitive or motor capacities. Those capacities must therefore be produced or stored in the affected parts of the brain. The only problem that spoils this line of thinking is the question: how does the brain do this? How does a process in the brain become a conscious experience? No one knows how the brain manages to do that. The only relationship between brain and consciousness that has in the meantime become known with certainty is that, conversely, consciousness changes and even forms the brain. We even know, down to the molecular details, how this formation of the brain, which is called plasticity, takes place.

Our Smart Brain

While I was standing there in the midst of the splinters of nineteenth century glass, and confronted with the raw reality of the consequences of 'we-are-our-brain-thinking,' it had already occurred to me that this way of thinking was probably not correct. Are we really saddled with a brain that rules us without us? Does this completely determine us? Or could it be that the window kicker was wrong? After a few decades of avid reading of neuroscientific books and articles in scientific journals, it has become clear to me that a gradual change has taken place in the way the role of the brain is viewed. The brain does not determine the human being, but the human being determines the brain. In a certain sense, the window kicker actually proved to have made, or at any rate, developed his own brain. And this can easily be compared with the way in which he managed to make his muscles so prominent: through vigorous exercise.

How does this work? The brain indeed provides us with awesome, biologically brilliant equipment that 'all by itself,' completely outside our consciousness, takes a lot of things off our hands so that we don't need to think about them anymore. Research by social psychologists has shown that when faced with decisions it is often better to rely on what our brains unconsciously arrange for us than on our conscious deliberations.[3] Conclusion: we should not stand in the way of our brain too much. 'Our brains are smarter than we are' could be a fair summary of the paradoxical results of such studies.

Plasticity of Connections

But how did the brain become so smart? Well, we did that ourselves. Of course, at first we have to work with what we receive at birth. But then things start happening. We have to use the brain, and in so doing we change the brain due to its plasticity. Naturally, we don't do that all by ourselves; we receive help, especially during our childhood years. Something happened in recent history that called attention to this. In Rumania, during the reign of the communist dictator Ceaucescu, a tragic, real-life experiment was conducted in orphanages, that demonstrated how important this is. The children were lying in cribs with high sides; they were fed by caretakers, but otherwise received no attention whatsoever. The result is shown in the following picture:

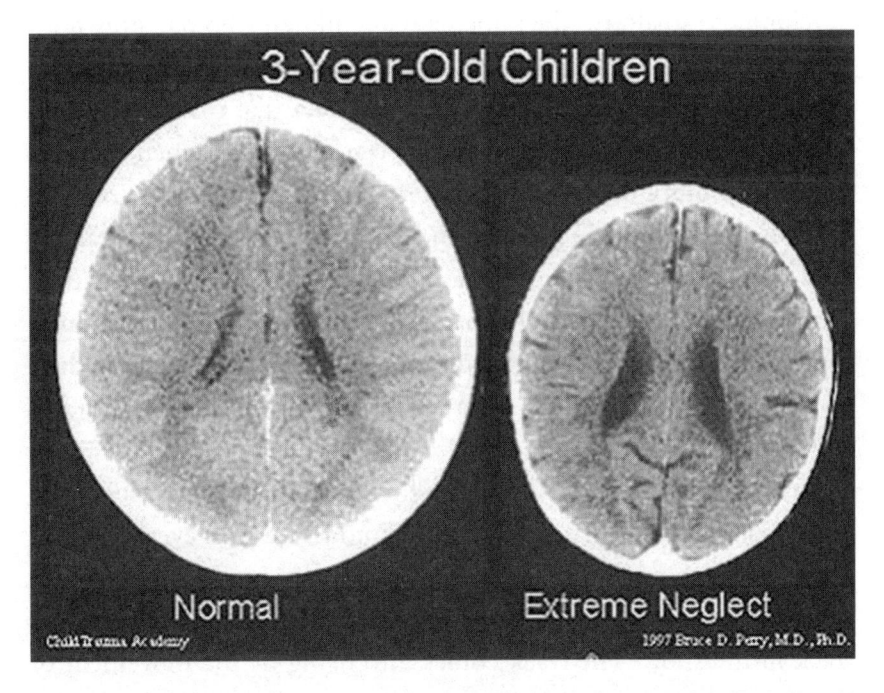

Figure 1. Brain development in normal three-year-old children (left) and under extreme neglect (right)

What we see here is not only that the brain has not grown, but also that the ventricles are enlarged, as are the furrows between the convolutions. They also show a notable deficiency in white matter (myelin) under the cortex, which is an indication of the connections and speed of thinking and, therefore, indicates intelligence. At the same time, the size of the skull shows clearly that the growth of the brain determines the growth of the skull. The children showed all kinds of stages of retardation, which was therefore not congenital but caused by neglect. The consciousness of these children had never been stimulated into existence by their caretakers. Conversely, much stimulating and loving contact with caretakers results in notable growth of the brain cells by giving rise to neurotrophic substances that allow nerve cells to grow.[4]

In 2015 an article appeared describing how a group of two year old children was taken from these orphanages and placed in foster homes. After about six years, they had caught up in their brain development to almost the same level as children who had grown up with their parents in the normal way. Children who had stayed in the orphanages did not develop further.[5] The fact that this latter group continued to exist as a control group seems cruel and unethical, but until this inquiry there were no foster families in Rumania.

Thus the window kicker was indeed a little bit right: we cannot hold someone responsible for the role of his parents and educators. On the other hand, it is also naïve to presume that a child comes into the world without any characteristics of its own; each in their own way, all children evoke a certain behaviour in their parents and educators.

How does the brain do that? The literature on the plasticity of the brain makes use of a generally accepted model for this: every thought leaves tracks in the brain in the form of nerve connections that form a network of their own. I can compare it with a cross-country ski track in the snow. The track comes into being by the movement of the skis. The path is made by the movement. Just as in the case of such a track, the next time we use the path it has become easier.

And as this takes place more and more often, it becomes a matter of course. Actually, it is quite logical that consciousness knows the way in the brain, for it is this same consciousness that has made the tracks, also the tracks of the unconscious.

This is the way we learn both knowledge and skills. When we fail to use the track for a long time, it snows over again and disappears. We are, however, creative beings, and we are perfectly capable of making a new track. Just as with a ski track this takes effort: consciousness, creative strength or will power. And the new track can also be deepened so that it results in a new network. This is the principle of plasticity.

We can formulate this process more neuroscientifically in two one-liners. The principles were already predicted by psychologist Donald O. Hebb, who suggested in 1949 that learning is based on new connections between neurons. When two neurons fire at the same time (synchronously), their connection becomes closer: 'Neurons that fire together wire together.' If they thereafter never do it again, the connection disappears: 'Neurons out of sync fail to link.'[6] Later this was simplified to: 'Use it or lose it.' Every thought, every activity of consciousness leaves its footprint in the brain. In the wiring language of neurology such a trace is called an 'open loop,' which is closed as soon as it is needed again.

Expertise

The brain, therefore, is partly formed by conscious and unconscious experiences after birth. The smartness of the unconscious has been brought into it by consciousness, voluntarily (meaning: out of free will) or not. It is like the story of the conductor who was to conduct the Boston Symphony Orchestra for its first rehearsal in Carnegie Hall, New York. He did not know the way and asked someone in the street: 'How do I get to Carnegie Hall?' The latter answered: 'Practice, practice, practice.'

Since the work of psychologist K. Anders Ericsson,[7] the conviction has grown that in order to become an expert in the fields of sports, arts or sciences it takes at least ten years, or ten times a thousand hours, of practice, practice and more practice. In so doing we change our brain connections in such a way that these enable us to automatically perform what we have practised, without having to consult our consciousness. It is therefore called the automation of the brain. According to Anders Ericsson, the most important characteristics of experts is ambition and the will to work. Experts have made their unconscious smart in their field. This automation works for all kinds of skills and capacities, whether actions or thinking. One could say that by exercise all forms of consciousness can be automated. But we can also learn and practise developing the brain without making any special effort and, as a result, increase our potential.

Speaking

The good news is that we are all experts in certain fields, for instance, in our mother tongue. We have practised this from birth, day in day out, much more than ten times a thousand hours, without any effort and with immense ambition and motivation to work on it. And we have been prepared for it even before birth. New-born babies recognise the voice of the mother and can distinguish their mother tongue (especially the vowels) from other languages; they even recognise stories they were told while they were still in the womb.[8] The time of childhood is the most favorable for developing these kinds of expertise, because there is then still an abundance of connections between nerve cells, of which the unused ones will be 'pruned' during adolescence.

This is the reason why, when we speak, we do not need to think about it first; the words come out all by themselves, whether or not perfectly formulated. The area in the brain we have developed in this process is called the language area, Broca's area (see Figure 2).

1. HOW COME THE BRAIN IS SO SMART?

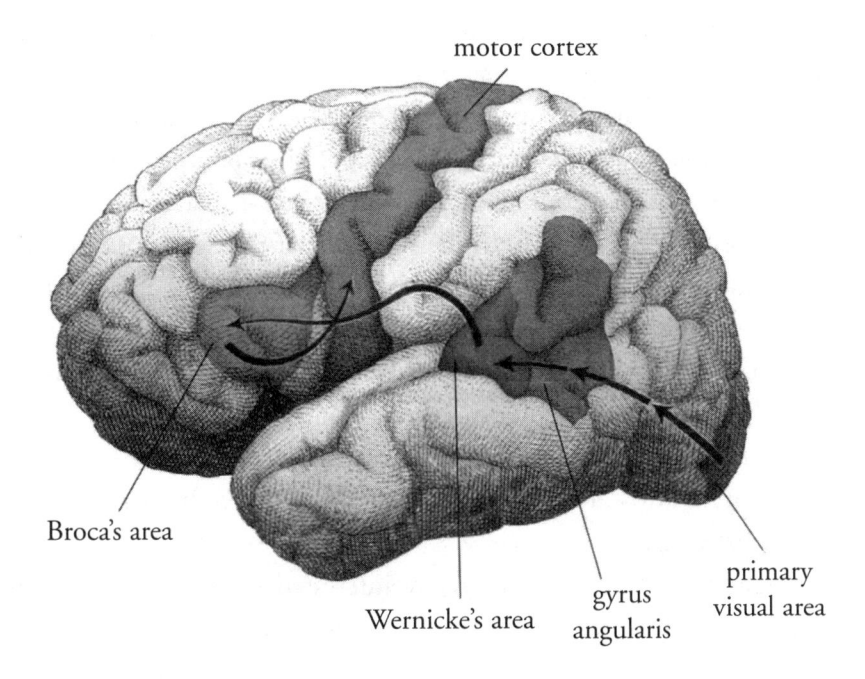

Figure 2. Areas of the brain

Broca's area is situated in the lower part of the motor cortex in the frontal lobe, and in monkeys it regulates only hand and mouth movements. From birth we have the tendency to imitate not only hand movements but also mouth movements and the corresponding sounds. Neuroscience connects this with mirror neurons which will be discussed in a subsequent chapter. Thus we develop Broca's area into one of the language areas, but it is equally the area of gestures.

The area that plays a role in understanding language is called Wernicke's area. The fact that we are able to develop the brain to the point that it plays a role in language, and that the areas of Broca and Wernicke do not even have to be actually present, is demonstrated by the examples of people who, from a young age, lack the left half of their brain which includes these two areas. These will be discussed in Chapter 15; as it turns out, they learn to speak in the normal way![9]

Faces

In the same way we become experts in the recognition of faces. There is an area inside the *gyrus temporalis inferior* called *gyrus fusiformis* (see Figure B on the inside front cover and Figure 6 on page 45, lowest convolution in top right picture) that performs additional analysis of what has come into the primary visual area in the occipital lobe (see Figures E and 6). In the right half of the brain a part of this *gyrus fusiformis* is called the area of face recognition (the *fusiform face area*, FFA). It is often described as an inborn capacity, ready for use in the brain. But this area is also one that we have to develop ourselves for face recognition. For this same area is also used for the recognition of a Maserati if you are interested in cars, a meadowlark if you are a birdwatcher, or a Modigliani if you are interested in the art of painting.[10]

Right from birth we are greatly interested in faces, even though a baby does not yet have sharp vision. And at a very young age we already recognise with the greatest ease the faces of the people we know. For we are still totally dependent on them. Human beings are experts in faces and have no need to look for the particular shape of a nose, mouth or eyes to know who the face they are seeing belongs to. Face recognition is of course immensely important, for we constantly read each other's faces and we experience that facial expression often represents someone's intention better than words. Interest lies at the basis of expertise. In this case, we have all from birth, without any effort, become experts in the field of faces.

Recognition of the faces of people living in other continents is often more difficult. Why do the Chinese or Japanese all look similar to westerners (and the reverse is also true)? Is this area of the brain then suddenly not working? And why do we become able to distinguish between these people once we are more familiar with them, and vice-versa?[11] Six month old babies can recognise faces of monkeys just as easily as those of humans.[12] That passes within the first year, unless the monkeys receive names.[13] It is therefore a process of identification of individuals, perhaps also a form of expertise.

The *gyrus fusiformis* is connected with a whole network in the temporal lobe (Figure E) to recognise faces.[14] Similarly, the right *gyrus temporalis superior* (upper convolution of the temporal lobe) is necessary to recognise emotions in faces.

According to neuroscientist Alva Noë, face blindness in its pure form does not occur all that often.[15] Could it be, when it has no neurological cause, that it might be a form of disinterest? In 2007 a research group in Miami found that unknown faces of people of whom it was said that they belonged to the same social or psychological group as the participant himself were better remembered than others.[16] Photographer Hans Aarsman relates in his book *De fotodetective* how, when male members of an isolated tribe in New Guinea saw six different photographs of the wife of a social development worker, they concluded in awe that the man had six wives.[17]

Similar to the language area, the face recognition area is not perfect and complete at birth like a module specialised in faces. In the left half of the brain the area homologous to the FFA on the right deals with reading written language. That can hardly be called inborn, even though there is in the brain of course always an aspect of inborn individual nature. Just as neurologist Oliver Sacks, who is face blind and wrote beautiful stories about his patients, cannot be accused of a lack of interest in people, some people can practise all their lives and still they will never overcome their dyslexia. There is a much larger area of the brain cortex that is dedicated to analysing what we see than just the area of face recognition. And all those areas have developed a preference for a specific aspect: sharp or obtuse angles, colour, vertical or horizontal movement, etc. A form of pattern recognition therefore. Because of this preference things may go wrong sometimes. We see then what we have learned to see, not what is really there; such perceptions are called visual illusions. One of the simplest is the Müller-Leyer illusion:

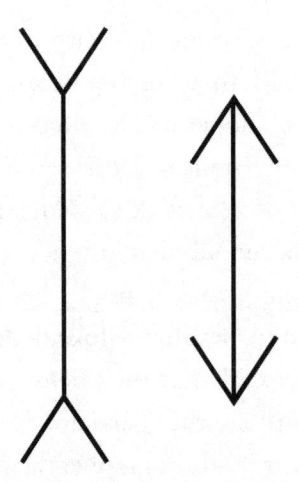

Figure 3. The Müller-Leyer illusion

I am sure you get the idea. The two vertical lines in the diagram are of equal length, but it is difficult to see that. You have to measure them to believe it. Now, the interesting thing is that although this illusion[18] trips up people in western industrialised countries, many peoples, especially those who live in small groups in huts, don't have this sort of problem at all, like the San in southern Africa, who used to be called Bushmen. They don't see any difference in length.[19] The presumption is that this has to do with the lack of straight lines and sharp corners in their living environment. We are used to translating two-dimensional pictures into three dimensions. When we see a photograph of a table, the surface is pictured in two dimensions like a trapezoid. Because we know that it is really a rectangle, that is what we see. To us the left line of the Müller-Leyer illusion seems farther away as if it were the corner of a room, while the line on the right seems nearer to us like the corner of a house. Our familiarity with perspective tells us that the left line has to be longer than the other one. That would have to mean that in a certain sense we have ourselves brought our illusions into our brain. The colours we see, and colour illusions, also prove to be caused by our experience.[20]

1. HOW COME THE BRAIN IS SO SMART?

Our Brains Become Who We Are

As is now generally known, the connections in our brain are not all genetically fixed. Throughout our lives, we can, by doing or thinking something new, change the expression of the genes – the task therefore of DNA – in the nuclei of the nerve cells, at least temporarily.[21] And every second we change the nerve connections in the brain; otherwise we would be unable to learn. At least up to our twenty-fifth year the brain, most of all the frontal lobe, develops under the direction of our own consciousness. And throughout our entire life we manufacture new neurons, from stem cells, albeit probably in only two places: the hippocampus (important for memory) and, to a lesser extent, the olfactory bulb. Recently neurogenesis has also been found in the striatum (basic nuclei such as nucleus caudatus, putamen and nucleus accumbens[22]) And in neurons even the DNA is changed by *transposons*, so-called 'jumping genes.'[23]

Thus intelligence turns out to depend not exclusively on heredity but also on what demands we make on the brain.[24] Intelligence is thought to be related mostly to the connections we have received at birth as well as those we produce ourselves (especially between the prefrontal cortex and the rest).[25] In brief, by our conscious experiences and activities we change our brain,[26] even our DNA. And in its turn, the brain then determines the scope of this automatic consciousness. Neuroscientist Joseph LeDoux expresses it subtly in the title of his book: *Synaptic Self: How Our Brains Become Who We Are*,[27] where he demonstrates that with our thoughts and experiences we change our brains, with the result that these 'coincide with ourselves.'

We ourselves are therefore the ones who make our brains so smart. We owe much of the knowledge we have regarding plasticity to a special family: the family Bach-y-Rita. That story follows in the next chapter.

CHAPTER 2

The Brain that Rewires Itself

The only model I had was how babies learn.
George Bach-Y-Rita

In 1959, at age 65, Pedro Bach-Y-Rita, a Spanish language professor at the University of New York and a poet, had a massive stroke. He could no longer move or speak. He was admitted to a hospital where he was kept in bed for four weeks, after which he was taken home in exactly the same condition. Then one of his two sons, George, a psychiatrist who was then still a medical student, brought him to his house in Mexico. Fortunately, George had not yet learned that the condition of his father was irreversible. To George it seemed that his father had to learn everything again from scratch the way a baby does. That meant: first crawling on all fours. In his all-male household – the father was a widower – they were not gentle and sweet to each other. George would, for instance, throw something on the floor and say: 'Dad, go get it!' There was a garden that father Pedro liked a lot, and George preferably let him crawl in that garden, to the horror of the neighbours.

But Pedro made progress and after some time was able to learn to walk erect supporting himself against a wall. He also learned to speak again, do the dishes, to type, and after three years was able to go back to work. He had command of all his functions again; he even remarried and worked for another five years. He died of a heart attack in 1965 at age 72 while climbing a mountain in Colombia at an altitude of 9,000 feet. Such a sensational recovery is not what we commonly see. The cases I know of such complete healing share one characteristic: they could develop thanks to the help of a loving family. The experience of being loved gives rise, as we saw in Chapter 1, to neurotrophic substances that allow nerve cells to grow.

But that is not the end of this remarkable story. The most surprising part is the following. The neuropathologist who did the autopsy published her report in the *American Journal of Physical Medicine & Rehabilitation*.[1] It turned out that the destruction the stroke had caused to the brain stem was still there, and was much greater than expected: 97% of the connections between the cerebral cortex and the spinal cord were permanently destroyed and, in addition, a large part of the motor cortex had died! None of this had recovered, so that Pedro's recovery must have been made possible by a reorganisation of the remaining functioning brain tissue. This is also the result of the plasticity of the brain. Driven by these events, Pedro's other son, Paul, a neurologist, specialised in neuroplasticity and revalidation of people with brain damage.

The above story can be found in *The Brain That Changes Itself*, a most interesting book by neuroscientist Norman Doidge.[2] The title is in conformity with the customary formulation in neuroscientific literature: *The brain rewires itself*. But is that in fact correct? Does the brain do that all by itself? Pedro Bach-Y-Rita came out of the hospital just as handicapped as when he went in. His brain really took no initiative whatsoever to change something, until George stepped in and took the matter in hand. The father was enormously stimulated by him and mobilised great persistence himself to change his brain. In so doing he did not repair his dead brain cells, but probably got the areas that have no special tasks, the so-called 'silent areas' where a neurosurgeon can remove tissue without consequences, to the point where they were able to shoulder new tasks. It seems to be such an obvious conclusion: when an area of the brain drops out and you can't move any longer, it has to be that area which makes movement possible. But this story also shows something else.

How can someone exercise something for which the brain area has dropped out? That would never be possible if that brain cortex were a prerequisite for that activity – if 'he himself were his brain.'

The secret of his recovery must therefore be looked for in 'himself' and not in his brain. The brain was his instrument and by playing on it again he repaired it, just as we constantly 'update' our instrument.

The Lollypop

Pedro's other son, neurologist Paul Bach-Y-Rita, made neuroplasticity his life work. He did experiments (he is no longer alive) on the use of senses that venture into the domains of other senses, such as touch impressions that engender visual experiences.

It has been known for a long time that blind people use their visual brain cortex (see Figure E on the inside back cover) to read Braille, not their touch cortex. Similarly, all the feeling in their fingertips and the touch that are transmitted by their cane is processed in the visual occipital cortex. And it does not take years to learn that. Seeing volunteers who are blindfolded for a few days and try to learn Braille very soon show the same activity in their visual brain cortex.[3] There is evidently no need at all for Ericsson's ten times a thousand hours of exercise for this. Sounds with which blind people orient themselves are also represented in the visual cortex.[4] Further, it has been shown that processing words that can be associated with pictures show visual brain activity in a homologous area for both seeing and blind people.[5] Therefore, it makes no great difference in what way information about the space around us comes in.

Paul Bach-Y-Rita developed a sort of lollypop with 400 stimulant electrodes, which is connected to a camera on the forehead. The prickles of the electrodes on the tongue function as 'pixels.' Because the tongue has a lot of nerve endings, when blind people put this lollypop on their tongue they learn in a short time to 'see' with their tongue.[6] In the BBC television programme *Focus*, presenter Michael Mosley demonstrated that, after practising blindfolded for only half an hour with this lollypop, he could throw darts on to a dartboard.

They all landed on the board, not far from the bull's eye. Quite an achievement, I think. Mosley reported that he experienced the tongue prickles virtually in the same way he would if he could see with his eyes. This achievement also turned out to be caused by the fact that the touch information of the tongue is processed in the visual cortex.

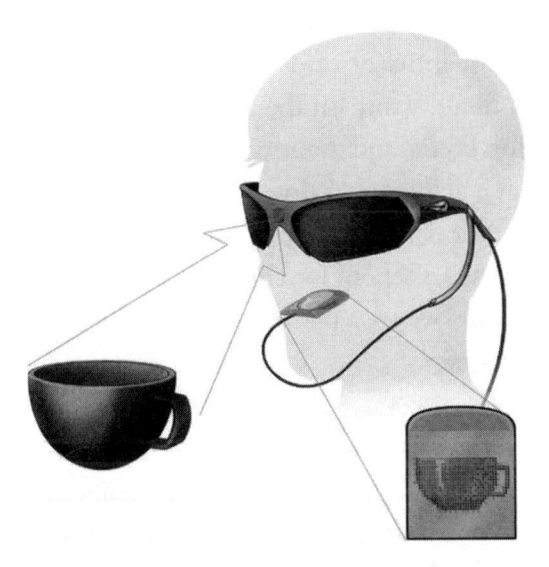

Figure 4. The lollypop of Paul Bach-y-Rita

So we see that areas in the brain can be mobilised with flexibility, a phenomenon known as the plasticity of the brain. As I have already mentioned, this is usually formulated as an autonomous activity of the brain itself. Does the brain indeed do this by itself? 'Who else?' might be the answer. And yet, we may have a strong suspicion that here again it takes place at the instigation of the consciousness and persistence of the owner.

The same is also true for the examples given by Doidge in his book cited above: the brain does not automatically go to work to change itself; it takes a lot of effort to bring that about. For instance, Doidge described a woman with learning difficulties. She could not understand metaphors and abstractions, did not understand the differences between above and below, right and left, and could

27

not tell time on the clock. The connection between the hour hand and the minute hand was a mystery to her. She could not follow conversations or dialogues in a film, and she had to read texts twenty times, and memorise them, before she could do anything with them. In this way she developed an incredibly good memory and managed to finish high school and go to college. She then read about neuroplasticity and embarked on a strict regime of exercises. She made a large quantity of cards showing a clock with a different time on each. A friend wrote on the back which time was indicated. She practised for hours and hours, practised and practised again, until she knew at a glance what time it showed. The surprising thing was that, in the process, her other handicaps also improved. In due course she established a school for children with learning difficulties. In her case, therefore, we see a person who, without help from others, changed her own brain.

The Brain Does Not Change By Itself

The plasticity of the brains of stroke patients is, in essence, an extension of our normal everyday plasticity: the changes in connections brought about by every one of our experiences. But that doesn't mean that the brain therefore changes all by itself. It takes more than that. How should the brain know which connections it has to make without our consciousness giving it direction? In the case cited above, such changes were made by the woman herself, because her consciousness managed to find its way through the brain and, in the process, made new connections. *We are therefore not determined by our brain, but adapt our brain to ourselves.*

But we don't do this all by ourselves, we need other people. But how do they change our brains?

CHAPTER 3

What are Mirror Neurons?

When people are free to do as they please, they usually imitate each other.

Eric Hoffer[1]

To understand the other, in other words, to reproduce his feeling in ourselves ... we generate in ourselves the feeling in accordance with the influence it has and shows on the other, by imitating with our body the expression of his eyes, his voice, his gait, his bearing (at least as a slight resemblance of the muscle movements and innervation). Then a similar feeling arises in us as a result of an old association of movement and awareness that leads to walking backward and forward.

Friedrich Nietzsche[2]

When my father finished watching the world championships ice-skating on TV he always had muscle pain in his thighs, although the only thing he did was to keep track of the statistics; he never liked sports himself. But he was a very good ice-skater in the winter, on natural ice.

In a cafe full of spectators of a soccer match you can sometimes see how someone's foot involuntarily shoots forward. Strange? Just try to be totally relaxed while you are watching someone making futile attempts to put the tip of a thread into the eye of a needle. Or try not to clear your throat while you listen to a hoarse speaker. But why? Why do the actions of others have such an influence on us?

Imitation

When you stick out your tongue to a new-born baby, the baby will right away imitate you. For young parents there may be no greater moment than the first time they see that the child imitates them, especially when it smiles for the first time. That is the moment when you know you have made contact. But it is also the first moment when you see that the child is learning something. For how does a child learn? How does it learn how to move, to speak, to express its feelings?

A child learns by imitating everything. That is what babies do all day long. It is the way we have made our brain as smart as it is: by imitating during our childhood. Of course this requires people to do things so the child can imitate them. It is really about interaction, just as in the examples given in the first chapter. Children learn by interaction, not by rules. As everyone should really know by now, in upbringing the good example is decisive, not the sermon. But how does a child know how it can imitate something? That is really a miracle. The child doesn't know that, it happens unconsciously, as it does for adults, by the way. For when someone shows you a movement you have never made, it is much easier to turn off your intellect when you try to imitate it, than to think about how to do it. We all remember that from when we were learning the sports we practise.

It has to do with one of the most fascinating structures in our brain, structures that according to neurologist V.S. Ramachandran[3] even determine our actual human nature: the network of mirror neurons and canonical neurons.

A Serendipitous Discovery

The story of the mirror neurons begins in 1995 in Parma, Italy, where scientists were doing research on lampong monkeys, members of the macaca family. They placed electrodes in their ventral (lowest part) premotor cortex to find out what the role of single neurons is in hand and mouth movements. This area, which plays a role in gripping, holding, pulling and bringing food to the mouth, is called F5 in monkeys. The story goes that someone came in who brought an ice-cream cone to his mouth and licked it. At that moment, the electrodes in the monkey of the day, while it was not moving, audibly registered activity in the area which until then had been regarded as the place in the brain cortex where only movements are initiated (and certainly not observed), specifically the movement with which food is brought to the mouth.

The ice-cream in the story turns out to be apocryphal but a comparable event did take place and is a nice example of serendipity, a completely unexpected discovery. Others might not have been struck by it, and the same thing may have occurred in another laboratory and may have been ignored as a flaw in the system, but this unlooked-for finding was neatly and without ulterior motives recorded in the lab journal. Only later was it noticed that this discovery did not fit in the expected outcomes which, therefore, appeared at first sight to be worthless. But slowly people began to suspect that they had made an epoch-making discovery. However, an article about it was rejected by *Nature* magazine due to 'lack of general interest.' After all, the question that was answered here had never been raised. In 1996 two other journals published articles on the subject in which the word 'mirror neuron' was coined.[4]

Aping

These articles made clear that there are areas in the brains of monkeys that not only play a role in initiating movements, but also in observing the same movements made by others. It was soon found that separate neurons were involved and not whole areas, and that they occurred not only in F5, but also in other areas such as the *inferior parietal lobe* (IPL), an area that American neuroscientist Ramachandran calls one of the specifically human ones (see Figure B on the inside front cover).

In human beings one cannot just stick electrodes into the brain (which may also raise questions of ethics in the case of monkeys), and therefore the expectation that human beings would also have mirror neurons could only be supported by Functional Magnetic Resonance Imaging (fMRI), a technique by which oxygenous blood flow, and therefore probably the activity, of various brain areas during various tasks, is shown in a picture. Single neurons could only be examined during brain surgery, when the brain lies open anyway, and only with permission of the patient.[5] Human beings were seen to have even many more mirror neurons than monkeys.

Mirror neurons are thus not only activated by the observation of an action, but also – just as clearly and in the same manner – by the execution of one. If that is true, it could have the consequence that we would constantly have to imitate each other. The reason that this does not happen lies in a soundly functioning frontal cortex (see Figure E, inside back cover), which helps us suppress that tendency. Areas that are developed only after birth (such as the medial frontal cortex lying on the 'inside' of the frontal half of the brain) have been shown to play a necessary role in this.[6] But still, it is not a complete suppression. When someone sees a movement made by someone else, activity has definitely been noted in the muscles of the onlooker. That explains the pain in my father's thighs; or the 'itch' you may feel in your hands when you see someone do something awkwardly which you can do very easily.

Besides mirror neurons that are activated in movement (in F5 in monkeys), other mirror neurons have been found that become involved only in purposeful actions. And even different ones (in the IPL) that fire at the mere surmise of the intention of an action. The action is then already mirrored although the person being observed has not even started it, or when only the sound of an action is heard, such as when paper is crumpled. In this way, even a utensil lying quietly on a table can call forth activity in neurons which are supposed to go into action when the utensil is used. For instance, a pair of spectacles activates neurons that come into action when someone puts them on his nose. Those 'action' neurons are called canonical. If these are not controlled because, for instance, the frontal cortex is damaged, the actions of others, no matter how embarrassing, are imitated without fail, or there may be an uncontrollable urge to try out every utensil in sight. Examples of this will be given in Chapter 10.

But healthy people are sometimes also defenseless in this regard. Think of a fit of the giggles; that this does not only happen with teenage girls is demonstrated by the YouTube film *Merci*, better known as *Bodhisattva in the Metro*, (which it is best to watch together with other people).[7]

Repeat after Me

What then are the consequences of the presence of these mirror brain cells? The area that is called F5 in monkeys is analogous to the area in the left brain of human beings known as Broca's area which, as mentioned earlier, plays a role in the formation of language (see Figure 2). Broca's area is packed with mirror neurons. This has thrown new light on the origin and the learning of language.

Linguist Noam Chomsky (born 1928) presumed that we have an inborn instinct for language. Neuroscientists expected this to be situated in Broca's and Wernicke's areas. But now it looks more as if we can learn language by imitating. This works only with attention and motivation. And those are things children have naturally, for they have a vital interest in learning to understand their parents or caretakers.

Both the area F5 in monkeys and Broca's area in humans play a role in mouth- and hand movements. This explains why speech and gestures have a strong connection with each other. This also makes it more understandable that sign language of deaf people, especially by imitating a speaker's body language, can be just as subtle as spoken language.

Mirror neurons, therefore, enable us not only to learn from our experiences, but also from each other, from each other's example. Without this possibility of transferring knowledge and skills by imitation, culture would never have been able to develop. To some extent, animals also have this capacity, but not in the same comprehensive and natural way. No wonder that Ramachandran considered these neurons essential for being human.

Empathy

Since even an intention of another person can be anticipated by the activation of the mirror neurons, the idea has arisen that the automatic and imperceptible imitation of a person's body language by someone else – even when not a muscle is seen to move – makes empathy possible. There is no better way to make the audience in a cinema cry than a weeping actor. We literally 'feel with' the other. For when we let our body speak through our muscles, the corresponding feelings are easily roused in others. Body language and feeling appear to be inextricably linked. Just try to be angry with a pencil clenched between your back-teeth while your lips must not touch the pencil.[8] Research has shown that muscle activity can be registered in a person who watches the face of someone in whom the same muscles are active. Similarly, when we look at a person who is laughing, the muscles we use when we laugh are activated, even if we manage to maintain a poker face. And this also puts us in a good mood (unless the person laughs at our expense).

Thus the position and movement of the entire body, including the melody of the voice, participates in conveying mood or emotion. And that is also mirrored, for instance, in the *gyrus marginalis* and *gyrus angularis*, which together form the IPL that in primates, and especially in humans, is strongly developed (see Figure B on the inside front cover). There are also indications that a talent for imitation goes together with a talent for empathy.[9] The reverse is also true. One of the characteristics of autism is a lack of ability to identify with another person. In the 1950's it was already discovered that autistic children cannot imitate others.[10] In the meantime it has been demonstrated by more than six different research teams, using a variety of techniques, that autistic children show disorders in the areas of mirror neurons.[11] With this, the cause of autism appeared to be evident. But later we will see that this was a premature conclusion.

3. WHAT ARE MIRROR NEURONS?

Sympathetic Pain
There are people who cringe when they see someone maltreated even if only on TV or in a movie. Most men are familiar with that when they see a man kneed in the groin. Can that be explained by 'motor' mirror neurons? Of course not. In the meantime it has been shown that mirror neurons not only occur in the two previously mentioned areas, but also in areas connected with physical or psychic pain or other feelings, namely the *insula* (hidden behind the fissure between the frontal and temporal lobes; see Figure E on the inside back cover).

It is suspected that many other areas in the brain cortex also have mirroring functions, for it has become evident that the mirror neuron network in humans is many times more extensive than in other primates.

Who Has Them and Who Doesn't?
Ramachandran may think that mirror neurons have made us human – not only because they make empathy possible, but also culture – but what then is the situation with monkeys where, after all, they were first discovered? Why did monkeys not become human? And might other animals also have these neurons?

Initially, mirror neurons were only found in monkeys, apes and humans. All three are great imitators, and apes and monkeys certainly demonstrate empathy. But there are other animals who show empathy, without being imitators. Invariably, these are social mammals and birds. On YouTube there is a video in which a dog rescues another dog that was hit by a car, by dragging it off the road.[12] Ravens console each other; geese get heart palpitations when their partner is attacked.[13] When our female goose had been killed by a fox, her mate loudly cried its heart out for three days. Geese also imitate; you can make clever use of that when you want them to go in a particular direction: with your fingers you make a goose head by laying your fingers close together with the fingertips touching each other. In that way you can send geese in any direction you want; they will follow your hand exactly.

Not much is known so far about empathy in the best known imitators among the birds. Parrots can precisely imitate sounds. I have seen a video of a bird of paradise that perfectly imitated the sound of a car door slammed shut and of the shutter of a camera. But some songbirds also learn each other's songs by imitating. And indeed, in one bird, a bunting, mirror neurons have been discovered.[14] Thus it would not surprise me if mirror neurons were discovered in other birds and mammals.

This means, therefore, that mirror neurons do not automatically make us human – animals also have them. Nor is the difference caused by the fact that we have many more of them. The point is that we have the freedom to obey them or ignore them.

Practising at Home
A logical consequence of the existence of mirror neurons would be that we could improve our range of movements by watching experts. This would hold true only for movements we have already practised so that the motor cortex has been prepared for them. In sports, dance and ballet, visualisation of movements without actually executing them has been used for ages before anyone knew of mirror neurons. Jan Jansen, a Dutch racing cyclist, went through the various laps of the Tour de France in his mind, and 'increased his efforts' on visualised upgrades. In revalidation situations, people are also told to visualise movements they cannot yet make very well.[15] It appears that in those cases mirror neurons are also used.[16]

Criticism
Criticism has rightly been voiced of the idea that we would derive empathy from a bunch of brain cells; or even, as Ramachandran suggests, that we owe our culture to these mirror cells. And the reader might well wonder what a chapter like this is doing in a book that wants to assert that we are not completely determined by our brain. Of course, mirror neurons do not produce empathy. Even if monkeys are not completely lacking in it, no one is particularly impressed with

any astonishing level of empathy and culture among them, and the bunting certainly has earned no fame for it. The point is that we human beings possess mirror neurons as an instrument, and that these enable us to use imitation for 'higher' purposes. But how do we get them? Suppose we are not (exclusively) determined by our brain, and therefore also by our mirror neurons, could it be that we possess the capacity to 'manufacture' mirror neurons ourselves? That they are simply the result of plasticity, as discussed in Chapter 1?

We Ourselves Do the Mirroring

That indeed turns out to be the case. *We ourselves also make mirror neurons.* For mirror neurons distinguish themselves in no way at all from other neurons, except in their connections. These enable them to fire both when you do something and when you see the same thing done by another. And according to current theory, the connections have probably come into existence because every time you do something, you also see and hear and feel that you are doing it. The relative neurons – those of perception and those of doing – fire simultaneously. And we know: 'Neurons that fire together wire together.' Those connections are reinforced every time, and that makes the relative neurons into mirror neurons or, rather, a mirror neuron network.[17] Therefore, here also consciousness directs the development of the brain, now specifically in the form of mirror neurons.

Since babies soon after birth imitate mouth and tongue movements, we assume that at birth they already possess some mirror neurons. They practise these movements already in the womb, sense the result and in this way prepare these mirror neurons. And according to neurologist Iacoboni, that is the beginning of the development of other mirror neurons: because the child imitates the parents, and the parents imitate the child again, and so on, this and other behaviour is imprinted in the brain and other mirror neurons are generated.[18] Thus mirror neurons are formed by interaction. The more interaction, the more mirror neurons. Before they are six months old, babies cannot yet imitate gripping movements. They must first learn to do them spontaneously, alone in their cradle, before they can imitate those movements. That is also true for learning to speak. The assumption is that by jabbering, a baby unconsciously associates the sounds it hears itself making with the motor program of the mouth and throat muscles it uses for those sounds. That produces a mirror neuron network that enables the child later to imitate the sounds made by the parents. As was abundantly demonstrated by the plasticity of the brain, here also: *the brain is formed in harmony with the way it is used, and this in turn enables us to make even better use of the brain.*

This continues into adulthood. Watching dancing ballerinas activates resonance in many more areas of the brains of other dancers than in those of people who do not dance. Conversely, the same was true for capoeira experts (a Brazilian fighting dance) when they were watching a capoeira match, to which ballet dancers reacted much less.[19]

Autism

In Groningen, Netherlands, a research team was investigating the connection between mirror neurons and autism.[20] Children with autism indeed showed less activation in mirror neuron areas. Contrary to expectation, it was demonstrated that adult autistic people did show activity of mirror neurons. This coincided with a higher degree of social adaptation. One of the researchers, experimental psychologist J. Bastiaansen, suggested therefore that autistic people do indeed possess mirror neurons but, at any rate in their childhood, do not use them in the usual manner.[21] Thus it may take a long time for an autistic child to learn to speak. It proves to be a question of attention. The attention of autistic people is different: when you talk with them, most of them do not look at your eyes, as normally happens, but at your mouth. Perhaps that means that they do not associate the words they hear with their own speech facility, but remain curious as to how the other does that.

This agrees with the theory that mirror neurons come into being when we observe what we do ourselves. But the consequence of this – the fact that these neurons also come into 'empathetic' action at the observation of others – is therefore not self-evident. Empathy is evidently not something that arises automatically in the brain. We have to want, and be able, to use the mirror neurons for it.

Use It Or Lose It?

Evidently, autistic people do indeed have mirror neurons, but don't use them. Later, if they have a mild form of autism, they can mobilise them by training themselves in imitation and empathy when they are confronted with their shortage of these qualities.[22]

Psychopaths have no empathy, according to the definition, but they are also masters in manipulation. To be good at manipulation, you have to be very sensitive and aware of what is going on in the other person. Indeed these people have plenty of mirror neurons, but it turns out that they are able to ignore[23] empathetic tendencies or must consciously engage them.[24]

The conclusion of this chapter, and of the previous one, is that not only do we ourselves (co)determine the quality of our brain, but that we are evidently capable of choosing whether or not to use the possibilities our brain is offering us. In the next chapter a woman comes on the stage who is able to turn part of her brain on and off in a radical manner. *This inevitably leads us to the question: who is in charge here, the brain or its owner?*

CHAPTER 4

The Instrument of Consciousness

*And he asked him: 'What is your name?' And he answered:
'My name is Legion, for we are many.'*

Mark 5:9

Around the beginning of this century a 33-year-old lady came into the office of psychiatrist B. Waldvogel in Munich, Germany.[1] She came with a guide dog for the blind because, after falling on her head fifteen years earlier, she had soon losst her sight. The reason she was seeing a psychiatrist was that she was suffering from dissociative identity syndrome (DIS), formerly known as multiple personality disorder. She showed some ten different personality states, 'alters,' or 'personae.' These proved to be of different ages, some of them were men, others women, some spoke German, others only understood and spoke English – she had grown up in England. Of course she had already seen an ophthalmologist who, after a very extensive examination with the most advanced technical means, had found no dysfunction in her eyes. That is quite possible for, after all, this was a case of psychiatry. She could have been 'hysterically blind,' which is a 'conversion disorder' due to which she would only imagine that she was blind. Conversion disorders are among the possible symptoms of DIS.

The article describing this event does not mention the cause of the dissociation, which could have been important to clarify the unconscious cause in the sense that there might have been something which the patient 'did not want to see.' We have all developed

4. THE INSTRUMENT OF CONSCIOUSNESS

defence mechanisms in our childhood in reaction to traumatic events in order to keep away unbearably painful experiences, feelings and memories. Dissociative syndrome is an extreme form of this. But for the significance of this case for the problem of seeing and the role of attention (that is, focused consciousness) this is not important, for an examination of the brain turned up something most interesting.

There is a test known as VEP (Visually Evoked Potentials). These potentials are mostly evoked by watching a chessboard pattern) that can measure electrical activity in the primary visual cortex, called V1 or *area striata*, in the back of the occipital lobe of the brain. This test can demonstrate whether or not the patient is able to see something, and if the result is positive, it would indicate that she was probably fooling everyone, consciously or unconsciously.

The remarkable thing was that this VEP test indicated no activity in her case. Her brain cortex did not react to light stimuli. Therefore, she really did see nothing, nor did she show any protective reflexes in bright light, or startle response (when threatened with being stuck in the eye by a finger), such as blinking, tears, closing the eyes, or showing fright. The diagnosis of the ophthalmologist therefore was: 'cortical blindness, probably the result of trauma to the skull.'

During the fourth year of her therapy – she continued to come with her guide dog – after a therapy session one of her alters could suddenly read a couple of words on the cover of a magazine. That was the first time in nineteen years! She only saw entire words; she could not distinguish separate letters. In subsequent sessions this alter also succeeded in seeing brightly lit objects, after which her vision soon expanded to its full range. In the course of further therapy the lady managed to get more of the other alters to the same point, but a few of them remained blind.

Had the brain cortex therefore now been restored? The next VEP test showed that for the alters that had remained blind nothing had changed. But the seeing personality states showed normal electrical response in the primary visual cortex. These alters could be called up simply by addressing them by their names. The transition from negative to positive electrical response was instantaneous.

Consciously Blind?

This story gives much food for thought. Psychiatry may perhaps not produce such trustworthy proof of the workings of our brain as 'hard' neurology with its examples of brain failure – and certainly a controversial diagnosis such as DIS seems at first sight inadequate to prove anything – but there is no doubt that we are facing an interesting neurological situation here. For what is really going on? What is the relationship between consciousness and the brain in this case?

In principle, there is nothing wrong with the visual brain connections of this patient. It is not a case of cortical blindness, for the visual cortex is capable of functioning normally. If there had really been cortical blindness followed by spontaneous recovery, which sometimes happens, the recovery would have had to take place shortly after the blindness occurred. But the ophthalmologist did not observe any variations in the cortical function.

In this sense, the patient was clearly not her brain. Would that it were so, since visually her brain was evidently completely normal. What is so interesting in this story is the fact that the visual cortex could instantly be switched on and off from the moment some of the alters could see again. Of course, that happens unconsciously. We can't just at will refuse the services of a brain region or allow it to function. Just try to be consciously blind.

And those alters are not consciously 'devised' in the moment. But evidently there is, at least during the therapy, a sort of director at work whose only task it is to hear which persona or alter is being addressed, and which therefore has to step forward and report its presence. But don't we all know this director? Doesn't he strongly resemble the director who lets us react appropriately to someone else? Isn't this the director who, when a policeman is about to write us a ticket, shows him a different version of ourselves than we show to our friends? Who is this director? Isn't that me, myself? In the above case, the woman's self was unfortunately not able to do this without the help of a therapist.

Both the blind and the seeing alters were lying ready in the brain in the form of their own individual network connections. At some point, in traumatic childhood, these connections were formed, after which they became 'blind' when the patient was eighteen years old. This was 'learned' unconsciously, just as all learning during childhood happens unconsciously. These kinds of dysfunctional defence mechanisms are the shadow side of the plasticity of the brain. In the case of DIS, some alters seem to be aware of the trauma, while others set themselves to avoiding it, and therefore develop amnesia or phobia of the trauma.

Fortunately, this same plasticity also creates the potential to change the connections of these alter networks through therapy, with the result that some of the alters gained access to the visual cortex again. But why some, and not some others? In therapy, it is all about consciousness. The way to overcome unconscious reaction patterns developed during childhood is to bring them to consciousness. Obsessive compulsive disorder, for instance, can be treated just by making unconscious behaviour conscious and by following conscious strategies. The result is then that changes take place in the brain.[2] Also in the case of depression we can observe the result of such therapy in brain scans.[3]

This is therefore another instance to show that the brain is changed through consciousness; and it is consciousness which 'succeeds in reaching' the visual cortex again.

Top-Down

The history of this blind and seeing lady is of course rather extreme. In her pathological case the switches from one persona to another, between the various states of consciousness, just happen unconsciously. But also for psychically healthy persons the state of consciousness, such as focused attention, is decisive for what they perceive. Directing our attention is something we do *ourselves*; we do it freely, to some extent at any rate, for our attention can also be drawn by something. But even then, we can never see something if we do not do the actual work through our consciousness, filled as it is with concepts, thoughts, feelings, will and memory, all of which serve to help us feel at home and know what we are doing in the world.

We can find the simplest example of this in the vase-and-face illusion, or in the cube that can be seen in two ways. What we see in these depends partly on our attention as conscious will. The visual cortex simply receives the signals of the pictures. These signals are unequivocal; they lack the ambivalence the pictures have for our conscious gaze. This immediately illustrates the difference between seeing and experiencing.

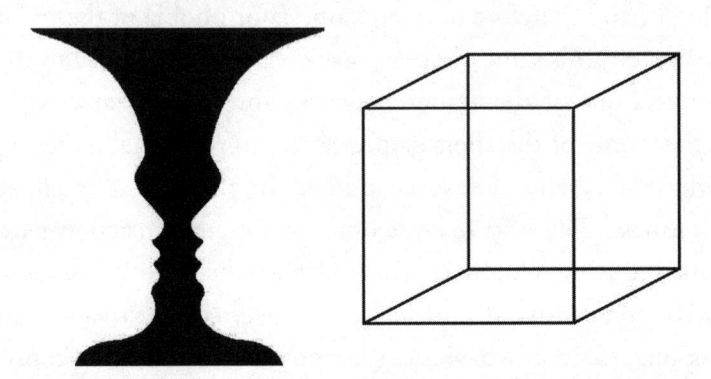

Figure 5.   The vase-and-face illusion, the two-way cube

There has been a lot of research into the effects of this free will in the visual cortex, in the form of attention, motivation and intention. It turns out that distinguishing objects can be recognised in a synchronisation of neuron activity among the several visual areas. It even appears that, when something becomes conscious, the relative neurons show synchronous activity.[4] This synchronous activity is visible in the EEG. Does the brain do all that by itself? Or does attention – focused consciousness – direct this synchronisation? Researchers have over time gained clarity on this question: it is attention and intention that take the initiative. It is called the top-down-effect, and the effect is significant, for the visual track in the brain does not end in the primary visual cortex, V1. The signals that come in there fan out to the front into the temporal lobe (see Figure 6) and up into the parietal lobe, to locations that are called association areas and are coded V2 through V5. In those locations the details of what appears in the field of vision are processed in the form of pattern recognition – colour, depth, form, movement – so that we can assign meaning to them. For instance, as

4. THE INSTRUMENT OF CONSCIOUSNESS

we have already seen, the face-recognition area lies just in front of V4 low inside the temporal lobe. Whether or not such an association area is switched on turns out to be a question of attention. According to researcher J.M. Hopf, the image on the retina evidently arrives in V1 without censorship, but we have, consciously or not, influence on what is done further with this image, and what ends up in our consciousness.

> These findings suggest that discriminative processes in the human occipital cortex ... are influenced by both the overall intention to perform a discrimination and by spatial attention.[5]

Of course, the researchers did not examine a lady with dissociative disorder who was both blind and seeing, otherwise they would have discovered that in such an extreme case their finding could even be true for V1.

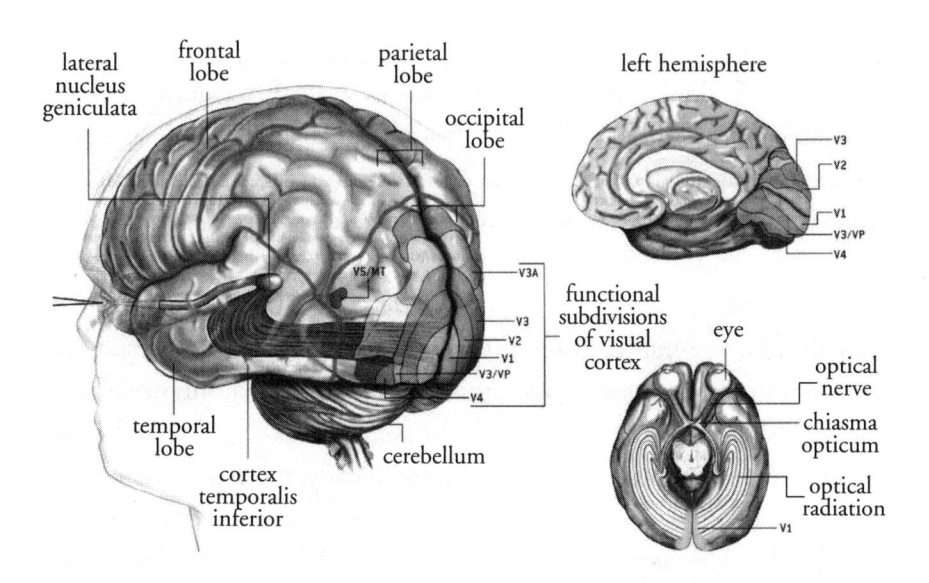

Figure 6. Visual track from eyes to visual cortex areas

45

It has even been shown that intensive attention or thoughts can influence the brain more strongly than the sense impressions themselves. Researchers showed photos of faces, animals or other objects to test subjects in whom, for clinical reasons, electrodes had been implanted that could register neurons separately. They found that for each of these pictures certain neurons became active – in the medial temporal lobe, for instance the *fusiform face area*. The researchers then presented two images on top of each other. The test subjects were now asked to strengthen one of the two images by intensively focusing their thoughts on it. This was possible because the electrodes were connected to the computer showing the images in such a way that the activity of a neuron determined which image was more clearly visible. And indeed, as the test subjects saw the image with increasing clarity, the corresponding neurons fired increasingly strongly, while the activity of the other neurons declined. Before this experiment, the assumption had been that we ourselves were unable to exert influence in that area of the brain.[6]

But the visual track from the eyes to V1 in the occipital cortex is not yet fully functional at birth; it has to be 'opened up' by consciousness. This begins already in the womb with the small amount of light available there.[7] When babies are born with a visual handicap such as congenital cataract, they need to have surgery quickly, otherwise the visual brain connection will never be able to function optimally.

Interaction

When new neural connections come into being as a result of consciousness, such as a thought, this does not mean that this new chain of connections is the thought. If that were the case, we would have to presume that these connections were already there beforehand, to make the thought possible. Suppose I read about smurfs who consume sarsaparilla. This activates a network in my brain. Does this mean that there is already a network of neurons present in the brain, ready for use, that occupies itself specifically with smurfs and sarsaparilla?

The brain is of course not static like that. In view of the fact that connections in the brain cortex constantly change, that can't be the case.

4. THE INSTRUMENT OF CONSCIOUSNESS

The thought arises undoubtedly at virtually the same moment that the connections change. We could thus wonder what is cause and what is effect. Are these new connections formed by new thoughts or, as maintained by neurodeterminists, are these new thoughts produced by the new connections? At this moment we do not know this, because the relationship between conscious experience and brain processes is obscure in both directions. But we can try to reason it through: those new connections clearly have to do with the new thought. Now, does the new thought adapt itself to the new connections, and does biology therefore determine what happens, namely, which thought arises? Or is it the nature of the thought that plays a decisive role here? Isn't it true that in a train of thought in logic, mathematics, philosophy and probably every other discipline, every thought would have to be the (logical) consequence of the previous thought? When we think, our thoughts usually follow from other thoughts. Would the changing neural connections do that too? Do they know logic?

Biologist J.B.S. Haldane, who coined the phrase 'primordial soup' for the beginning state of life on earth, clearly recognised this problem of the inner relationship of thoughts versus the statistical probability that reigns in the world of material atoms and molecules. That was in 1932, and it may seem remarkable for a declared materialist:

> It seems to me immensely unlikely that mind is a mere by-product of matter. For if my mental processes are determined wholly by the motions of atoms in my brain I have no reason to suppose that my beliefs are true. They may be sound chemically, but that does not make them sound logically. And hence 1 have no reason for supposing my brain to be composed of atoms.[8]

Now, in the brain we do not witness accidental movements of atoms such as in lifeless nature, but purposeful molecular processes of biology. Therefore, the production of logical thoughts would mean that logic, meaning and truth are imprinted in a purposeful biological process.

This is what we would have to believe in order to assume that the brain produces thoughts and consciousness. If not, it is obvious that we ourselves use the brain as an instrument – call it a brilliant biological computer, a computer we not only use, we also program it.

The supposition that consciousness is not simply a product of the brain, but that we make use of the brain as an instrument for consciousness, is something which, in view of the examples of plasticity in the foregoing chapters, where consciousness also played a decisive role, I find much more credible than the reverse. Our brain does not think; we think with our brain, just as our legs do not walk but we walk with our legs. Thomas Fuchs, Professor of Psychiatry in Heidelberg, Germany, expressed it as follows:

> This reciprocal relationship of 'process' and 'structure', with each of the two poles enabling and modifying the other, is the foundation of the joint development of mind and brain. It also strongly contradicts any reductionist notions of the brain as the creator of the mind.[9]

The picture that begins to emerge of the relationship between the brain and consciousness is like this: Consciousness changes the possibilities of the brain with the intention that the brain, as an instrument, then lets consciousness make use of these possibilities. A connection, once created, makes it easier to send a certain thought or behaviour in a certain direction. This is what learning signifies, and it is also the process that makes it possible to develop an automatic skill. Consciousness and the brain interact with each other.

Consciousness and the brain keep each other in balance. Consciousness forms the brain, but this only makes sense if the brain has a task: to serve and attend to the needs of consciousness and behaviour.

In the case of the blind and seeing patient, the problem was that, by the strategy she chose to deal with the traumas experienced in her youth, she developed a personality disorder that was so deeply anchored in her brain that she indeed 'seemed to have become her brain.'

4. THE INSTRUMENT OF CONSCIOUSNESS

But there are two reasons to observe that she is more than her brain. The first is that she caused it all herself, albeit involuntarily. And the second reason is that the psychiatrist was able to mobilise something in her (the 'director,' her *self*?) that made it possible to put things in order again and bring the condition to an end.

Similarly, there are many more conditions of which one could say that these people 'are their brain,' such as psychoses, hallucinations, compulsive thoughts, addictions, depression, in brief, all those psychiatric conditions of which you get the feeling that the person has lost control, and has therefore become unfree. And that is probably exactly the right description and potential cause of these conditions: incapacity, temporary or otherwise, of the *self*. Indeed, therapy is possible. And it consists, apart from potential medication, precisely in the mobilisation of the *self* so it can become present again, as the case of the blind and seeing lady shows. In a certain sense, her *self* was then present (again) in the visual cortex.

By the way, it needs no psychiatric diagnosis to find ourselves in such a situation. We are all familiar with these kinds of conditions to a greater or lesser degree. We are not always equally strongly present *ourselves* in what we do.

CHAPTER 5

How the Pictures are Interpreted

We can carefully study the structure of the brain and say: if we think, this and that happens, but what thinking is, is something we know better through ourselves.
                                                    Frederik van Eeden[1]

When thirty years ago we wanted to make a picture of a living brain, we made X-rays of the brain and, for contrast, air was injected into the ventricles (pneumencephalogram). It was an extremely painful procedure that yielded very little information. Except for the inventors of the first scanning machines, no one then thought it would be possible to create an image of living brains and brain activity. At that time, only the EEG – which could not show any precise location of activity – and the post-mortem gave information on potential shortcomings of the brain. In the past twenty years there has been enormous progress in the quality of brain examination, especially due to the fMRI scan (functional MRI, a scan image of brain *activity*). This is a wonderful development that has resulted in incredible insights into the working of the brain. And it has not been hidden from view, but has been amply reported in the media.

All this media attention has had an interesting consequence: statements about psychological phenomena are apparently more convincing when they are accompanied by neuroscientific information, whether relevant or not. The intelligence or level of education of the reader makes no difference, unless he or she has been trained in the neurosciences.[2] The effect is even more pronounced when pictures of brain scans are included in the articles.[3] The mere fact that an article deals with neuroscience, especially if it has brain images, makes it believable. (And of course this book also benefits from it.) What Marshall McLuhan once asserted about the electronic media – 'The medium is the message' – also holds true for the neurosciences: *A statement with a reference to neuroscience is convincing.* And such statements can be quite extreme.

Reading Thoughts

Neurodeterminists such as Victor Lamme, cognitive neuroscientist at Amsterdam University, are convinced that an fMRI scan can enable an examiner to know better what a person is thinking than the person himself. Thus, according to Lamme, a personal, subjective experience can definitely be examined from outside. When a person in the scanner is shown pretty, neutral or unpleasant pictures, one can see by the activity of each little piece of the brain – 3×3×3 millimetre called a *voxel* (volume pixel) – whether that little piece 'likes' positive or negative pictures.[4] When in such a case the person is shown pictures of cars, the balance between 'pleasant' and 'less pleasant' voxels tells the examiner better than the person himself which car he would like to buy. Such knowledge would be invaluable for business. It has led to the development of a new specialty: neuromarketing.

It is a little ironic, to say the least, that the examiner would know better what someone thinks than the person himself. For this leads to endless regression. Suppose that the thoughts of the examiner would also be made visible by an fMRI, then the observer of that brain scan would also know better what the examiner thinks than the examiner himself, and so on. And this leads us far away from the first person's own thoughts. It is an idea that can only occur within the view that the brain is the producer and seat of consciousness. Consciousness is in this way separated from the body and its environment and can be studied from outside from the perspective of a third person. But the examiner himself is someone with the perspective of a first person.

One can never ever leave the personal point of view out of any science. There is always someone who interprets. To deny this is the point of view of a behaviourist. And Lamme indeed calls himself a behaviourist.

Perhaps a brief explanation is due here. Behaviourism is a trend in psychology that says that a person's thoughts and feelings are of no importance in a psychological examination. According to a behaviourist, what a person says of himself does not produce reliable information; neither does introspection as practised by Freud. What counts is the behaviour a person shows. That is the only thing that demonstrates what a person wants, not what goes on in the soul. This means that psychological examination can just as well be performed on animals. The focus is then on stimulus, response and conditioning, with which behaviour can be changed. The stimulus will take the form of award or penalty. Since the 1950's this trend has no longer been taken seriously by psychologists. It was replaced by cognitive psychology in which mental phenomena are considered important.

The best way to show the consequence of behaviourism is the dialog between two behaviourists who just had sex with each other: 'Wow, that was clearly fantastic for you. Tell me, how was it for me?'

In this view, the activities visible in an fMRI scan are seen as behaviour. But what are we to think of the idea that the voxels in the fMRI scan really precisely show what a person thinks? Is that at all possible? Are thoughts so clearly unambiguous? Is this not the same kind of reasoning as the opinion that we can receive the message of a movie by studying the pixels of a DVD on the screen?

It would open up tremendous possibilities. The police or a judge could check whether a suspect was lying. Unfortunately, a test has shown that suspects can easily frustrate this procedure when lying in the scanner by simply making imperceptible little movements with their fingers and toes.[5]

A different, legal consequence of the idea that the brain truly shows someone's thoughts and feelings, as opposed to what the subject himself thinks, is called in neuroscientific literature the 'my brain made me do it' effect. It does not quite mean 'we are our brain,' but more 'I did not notice what my brain made me do; therefore I can't do anything about it.' It is the point of view of the window kicker in Chapter 1, which until recently was only applied to animals. By definition, animals cannot do anything about what they are doing. It is interesting that this argument is only used in cases of unpleasant things. No one will advance it to excuse himself for doing a good deed.

'How to Explain Pictures to a Dead Hare'

Neurodeterminists thus postulate that consciousness is a thing that can be made visible. The Dutch psychiatrist and author Frederik van Eeden continued the quotation at the beginning of this chapter as follows:

> We could know exactly what takes place when someone is in pain, what changes in his nerve elements; we could discover the anatomical and physiological substrata of pain. But someone who has never experienced pain will not sufficiently learn from this what pain is.[6]

Doubt as to the idea that an examination from the outside is really able to say something about someone's inner experiences led neuroscientist Craig Bennett to a remarkable experiment. In 2009 he laid a salmon in an fMRI scanner and showed the animal the usual pictures. In reaction to several pictures the salmon brain actually lit up in a number of places. We should of course not underestimate salmons, but this animal had been dead for at least two days.[7]

It made me think of the German artist Joseph Beuys who in 1965 gave a performance with the title *How to Explain Pictures to a Dead Hare*.[8] It was intended as an ironic commentary on the baroque prose with which art critics manage to decipher the most fantastic intentions in the art works they have to review.

There is evidently quite a lot of static in fMRI's. An article published in 2016 reveals that since its invention fMRI tests have produced thousands of unreliable results. Three commonly used software packages were tested that are responsible for these results. As it turned out, up to 70% of positive results shown were incorrect.[9] However, the problem does not lie in the as-yet imperfect technology, but rather in the ideas behind the design of the experiments.

My Inner World in a Picture?

If Bennett had taken a living salmon he could also have demonstrated something else. For the purpose, we should then have to assume that a salmon is interested in pictures. If the fMRI test had shown activity in the brain of the salmon, and we know with which structures of the human brain these locations correspond, would we then know something about the inner life of the salmon? Would we then know what it is like to be a salmon? No, of course not. (This argument was proposed by the philosopher Thomas Nagel in his article *What it is Like to Be a Bat*, in which he states that, even if we know all about a bat and the way it orients itself with echolocation, we will know that only as human beings, and not the way a bat experiences it.)

Why then do we think that we do know this in the case of human beings? In the end, we cannot know more than what people have first communicated while looking at 'positive and negative pictures.' But even then we do not know how these are inwardly experienced by them, what their *first-person-perspective* is like. Feelings and thoughts are not two-dimensional like fMRI pictures, or three-dimensional like voxels. They have no dimensions.

But before entering into the details of this critical point, I want to mention a few technical difficulties that might cause us to take the results of fMRI tests with a grain of salt.

- fMRI is much slower than our inner experiences, but that will most likely improve in the future.
- There is no exclusive, one-on-one relationship between brain locations and mental processes. Most, if not all, brain areas are involved in multiple mental states.

5. HOW THE PICTURES ARE INTERPRETED

- When the press publishes another report on the spot in the brain that is active in love, or in the experience of beauty or in all kinds of other experiences, we always see the same structures appear like celebrities in a glamour magazine with the probably less well known names: medial prefrontal cortex, anterior cingular cortex, nucleus accumbens, caudate nucleus and amygdala. Especially the medial prefrontal cortex (on the inside of the frontal lobe) often appears in fMRI pictures. That is obvious, will be a neuroscientist's answer, because this area is considered to be the centr of so-called executive functions, the functions we need when it is important that we stay in focus, such as in planning and decision making, correcting errors, new behaviour and breaking habits, and in dangerous situations. But it is a rather extended area in which these voxels may light up.[10]
- During the extremely stifling situation in a scanner (head immobilised, loud sounds, small space) pictures are supposed to represent situations in ordinary life. How reliable is that?
- What we see in pictures of fMRI scans is the result of statistics: only the areas in which the activity exceeds a certain minimum level are used by the computer to show them as a specific area, in a particular colour. It is quite possible that areas that are important do not show up, and the reverse. Small statistical differences may be exaggerated in colour differences. In addition, there is the problem that the convolutions in the brain are different for everyone, as are the locations ascribed to various functions.[11] The results are averaged for the different test subjects, so that an area may be very active in one person and not at all in another. This was observed, for instance, in a test of the effectiveness of acupuncture, notably in the control group that did not get acupuncture but only had to drum with their fingers. When the results of this

drumming control group were examined, it turned out that the same test subjects always showed different results in different sessions in the fMRI while they were drumming in the same way.[12] Therefore, even something as simple as drumming with the fingers either is not located in the same brain areas all of the times or the fMRI scans are unable to show it.

- The 'average location' is therefore an artifact. In the example of Lamme, however, this is not important, for there each participant is individually gauged, so that the functions of the different areas are separately observed for each.

- In one test the participants were instructed to simulate a particular personality – adaptive or its opposite, extroverted – while lying in the fMRI scanner. This role-playing had an influence on the result. When extroversion was simulated, the result of the scan showed just that.[13] Although the title of the article describing this is *How the Brain Creates and Uses Personality Models to Predict Behaviour*, we might well wonder who does this: the test subject or the brain.

All these difficulties lead to the really serious problem that was noticed by Vu and Pashler. In most of the fifty-five articles they examined about brain locations found by fMRI scans, including those that had appeared in the leading journals *Nature* and *Science*, they 'show that these correlations often exceed what is statistically possible, assuming the (evidently rather limited) reliability of both fMRI and personality/emotion measurements. The implausibly high correlations are all the more puzzling because social-neuroscience method sections in these journals rarely contain sufficient detail to ascertain how these correlations were obtained.'

And this had to do with the usual statistical computer procedures.[14] In brief, the results were many times too good to be true, and there was enough reason to doubt that the active areas and the emotions really had anything to do with each other. All of this should urge us to be most careful when drawing causal conclusions from neuroscientific imaging techniques.

But none of these forms of unreliability address the principal difficulty I want to discuss. The really critical problem, as far as I am concerned, is the reduction of complex feelings to fMRI pictures. This is not just a technical, but also a philosophical sin, a categorical error inspired by the materialistic idea that consciousness is identical to bodily tissue: the brain. Just as an ophthalmologist no longer sees my gaze when they examine my retina, a neurobiologist does not see my feelings in the brain scan. At best, when they see my body language, physical position and movement they can, with the aid of their mirror neurons, sympathise with what I feel. We can find many examples of this short-circuit.

For instance, researcher Mario Beauregard, who takes a special and sincere interest in 'higher' feelings such as altruism and religiosity, found the brain area where the neural basis of 'unconditional love' is located. He found it by showing pictures of normal and handicapped persons to caretakers of handicapped people, and asking them to feel unconditional love while looking at these.[15] What became active when they were looking at photos of the handicapped had to be the centre of unconditional love. He described eight different areas that became active. Semir Zeki found the places (indeed different ones from those found by Beauregard) where beauty is processed by showing people pictures they had classified beforehand as pretty or ugly.[16] And Zeki and Bartels found the 'neural basis of romantic love' by showing pictures of neutral faces and the face of the loved one of the test person, who at that moment had to be terribly in love.[17] (A few areas were the same as for Beauregard's unconditional love: the *medial insula*, the *cingulate cortex anterior* and the *caudate nucleus*) Neuroscientist Helen Fisher later also found where 'romantic love' can be found, similarly by having people think of their loved ones. And indeed, there the *caudate nucleus* appeared again.[18]

Geriatric neurologist Raymond Tallis calls all of this '...a crude experimental design which treats individuals as passive respondents to stimuli and then discovers that they are passive respondents to stimuli.' And, he adds:

> As anyone knows who has been in love – indeed anyone who is not a Martian – love is not like a response to a simple stimulus such as a picture. It is not even a single enduring state, like being cold. It encompasses many things, including: not feeling in love at that moment; hunger; indifference; delight; wanting to be kind; wanting to impress; worrying over the logistics of meetings; lust; awe; surprise; joy; guilt; anger; jealousy; imagining conversations or events; speculating what the loved one is doing when one is not there; and so on.[19]

Does fMRI therefore enable an examiner to see whether a test subject loves someone? And to know that better than the subject himself? Undoubtedly, fMRI pictures have something to do with the soul activity of the subject. But what exactly? An fMRI examination takes no account of the fact that we have an embodied consciousness that cannot be captured in a brain scan. That has to be strikingly evident, for instance, in love. What brain scan will show the butterflies in the stomach? Or the dreamy gaze, the uncertainty or the desire? Consciousness is a much more encompassing phenomenon than firing neurons. Reality is much too unruly, feelings too complex, and the variety among individuals too great to draw such simple conclusions. Moreover, the examiner sees the exterior, while the subject experiences the interior, two totally different categories. After all, the location of a yellow or red spot in the fMRI picture in no way enables the examiner to experience the quality of the subject's supposed love, no matter how much expertise he or she has developed.

What Does an fMRI Picture Actually Show?

What is this exterior anyway? What causes a particular brain area to become active? What prompts a neuron to fire? The answer is: the action potential of another neuron, or better, the neurotransmitters originating in another firing neuron. This kind of reasoning just changes the question: at the end of the day, what is the cause of the chain of phenomena? Here we find ourselves in a vicious circle. The neurodeterminists presume that the brain does all this on its own authority. But where does this autonomous brain decide that in one case it will not participate, so that these neurons will not fire but perhaps others will? And how does the brain decide which task it will embrace, which brain area can go to work?

We started from the premise that an fMRI, as image-forming technology in neuroscience, can demonstrate which brain areas are doing the work. But what does an fMRI scan actually show? In its most commonly used form, BOLD-fMRI (blood-oxygen-level dependent), the scan measures the increase in blood flow. A change in magnetism occurs when the ratio between fresh, oxygen-rich and used, oxygen-poor blood is increased. The areas with fresh, oxygen-rich blood are shown, after computer processing, in bright colours, from yellow to red. Because the variations in oxygen content take much longer than neuron activity, we don't really know in which sequence this all happens. First brain activity and then blood flow? Or the other way around?

What an fMRI shows, therefore, is not a direct image of neuron activity, but the increase in blood flow. This was never a problem, because until recently it was assumed that these things meant the same thing: active neurons simply demand more blood. This is taken care of by astrocytes, which could detect neuron activity.[20] (Astrocytes are brain cells that are not nerve cells, but so-called glial cells that take care of a large number of auxiliary tasks. Possibly, because of their relationship to neurotransmitters, they might also play a role of their own in consciousness and psychiatric disorders. We have about as many of them as neurons.)

However, two researchers wondered whether this was actually true. They timed the events with great precision and discovered that blood flow increases before there is any demonstrable activity of the neurons in the relative area.[21]

In the first place, this would signify that fMRI images need not necessarily coincide with neuronal activity and, secondly, perhaps it is not the activity of the neurons that attracts blood, but the other way around: that the blood flow determines where neuronal activity takes place. Blood flow would then activate neurons. But who or what sends the blood there? When we start looking for answers to such questions, for the beginning of the causal chain, we can merely come up with hypotheses. These can be either deterministic – 'the brain does that, like a reflex, probably in the prefrontal cortex' – or non-deterministic – 'we do not know this, but could this be something that points to a 'self'?' The latter idea is somewhat speculative because this research has been performed on animals. Do animals have a self?

The question of what initiates brain activity is of equal importance to the question of the relationship between consciousness and the brain, but perhaps even more difficult to answer. What sees to it that not one part of the brain but another part is put to work? Does that admit of a causal and deterministic explanation? Many authors of brain books think it does: it is simply a question of reflexes. Everything takes place within the head.

It will be a recurring question in this book: Does the brain determine consciousness or is it the other way around? We might even wonder whether consciousness is actually located in the head.

CHAPTER 6

Where Does Consciousness Reside?

We are empty heads open onto a single evident world.
Maurice Merleau-Ponty[1]

The BBC has shown a number of programmes about the brain. In one of these, *Focus*, Michael Mosley, who is a former general physician, presented an instalment on pain, in which he had his whole body waxed to remove the hair, in an effort to keep up with the times. As the hair on his calf was removed with a quick, cruel pull by a determined lady, he cried out in pain and immediately gave the following commentary: 'It may seem as if I feel pain in my calf, but in reality of course I feel it in my brain.' However, it is rather interesting in this regard, that no pain is felt in brain tissue. A neurosurgeon can cut into the brain while the patient is fully conscious and feels nothing. Thus whatever may occur in the brain, we do not *feel* anything there. While watching this program I wondered whether Mosley, when he still practised as a physician and asked his patient 'Where does it hurt?' would have corrected the latter's answer by saying: 'You may think so, but actually…'

The theoretical model of pain tracts that situates the experience of pain in the brain is therefore more real for Mosley than his bodily experience. What is the story here? Where do we feel something? Where does the consciousness of pain reside? In the spot where we experience it or in the brain? Or, where do we scratch when we feel an itch?

Pain Is Consciousness

Obviously, we need the brain to feel pain. Most pain medication works on the brain. And that could mean that the feeling of pain is manufactured there. But there are rather convincing indications that in this case also it is not the brain that determines what we feel or perceive, but that it is consciousness that determines the processes taking place in the brain.

First of all, we have pain only when we are conscious. When we sleep, we feel no pain. Pain can indeed wake us up, just as sound or light may wake us up. And during REM sleep, pain can influence dreams. It is interesting that when pain is inflicted during sleep, the centres that are connected with the localisation of pain are activated in the same way as when this happens during waking consciousness. This was found when pain was caused in the skin of the hand of a test subject by a laser beam, both when he was asleep and awake. The only difference was that during REM sleep the middle of the *cingulate cortex* or *gyrus* (Figure D on the inside back cover), which has to do with orientation and avoidance behaviour, was not activated.[2] (During REM sleep the body is, in a certain sense, paralysed so that we do not make the movements that are made in the dream.) But otherwise the reaction in the brain to the pain stimulus was the same while awake or asleep.

Thus there is an unclear relationship between consciousness of pain and activation of the brain cortex. For instance, what is cause here and what is effect? Most people, and most neuroscientists, think that brain activity produces consciousness, but there are others who consider it possible that it is the other way around. How do we find out about this? If it were possible to trace a sequence in time, the causality would be evident. In the 1960's neuroscientist Benjamin Libet did some remarkable research into consciousness of a pain stimulus and its processing in the brain. What this experiment shows should cause us grave doubt as to the idea that consciousness resides in the brain. The neuroscientific community therefore did not know what to do with it, and the experiment is hardly ever quoted.

6. WHERE DOES CONSCIOUSNESS RESIDE?

Libet showed that a stimulus applied directly to the skin is felt more quickly than when the brain cortex that belongs to that area of the skin is stimulated. He did the following experiment: into the brains of test subjects who underwent surgery by Bertram Feinstein because of Parkinson's disease or incurable pain, he implanted electrodes in the spot in the sensory cortex of the brain that corresponds with the back of the hand. When a series of electrical stimuli was applied to this spot in the brain cortex, the subjects felt these as if they were applied to the back of the hand. It was the same feeling as if the back of the hand was given electrical shocks directly. However, when the stimulus in the brain cortex lasted only a very short time, less than half a second, the subjects never noticed it at all. They only became conscious of the stimulus if it lasted more than these 500 milliseconds. By itself this is not so shocking, and it appears to support the view that the brain produces feeling. But now note the following: Libet then compared the stimulus to the brain cortex with a single stimulus to the back of the hand. To do this he first stimulated the spot in the brain cortex corresponding to the back of the right hand, and 200 milliseconds later the back of the left hand directly on the skin. What should we expect if we are convinced that feeling is manufactured in the brain? We would obviously expect that the stimulus on the left hand was felt at least 200 milliseconds *after* the one to the brain (which was felt on the back of the right hand); and most probably even later than that, because the former had to cover the whole distance from the hand to the brain. But this was not the case. The stimulus applied directly to the left hand was felt the instant it was given; the one to the brain cortex corresponding to the back of the right hand only after 500 milliseconds, 300 milliseconds therefore after the one to the left hand while it had been applied 200 milliseconds before!

Naturally, that is strange if one is convinced that consciousness only arises in the brain. For that would mean that a stimulus in the cortex has to be felt more quickly than one on the skin of the hand, since the latter still has to travel to the brain. Libet came up with a near-supernatural solution for this. He suggested that the feeling was projected from the brain to the hand (after all, the stimulus in the brain cortex was felt on the back of the hand) and simultaneously also experienced a regression in time. He formulated it as follows:

> …there is a subjective referral of the sensory experience backwards in time. … The specific projection system is already regarded as the provider of localised cerebral signals that function in fine spatial discrimination, including the subjective referral of sensory experiences in space. Our present hypothesis expands the role for this system to include a function in the temporal dimension.[3]

But how should we picture that? Both forms of projections go against natural laws. The only reason he has for suggesting such a far-fetched solution is the assumption that consciousness arises only in the brain.

Scientists often like to refer to Occam's razor, so named after a medieval English monk who said that of all the possible solutions to a scientific problem the simplest one has the best chance of being correct. In my view, there is a much simpler – and only seemingly naïve – solution possible than the one suggested by Libet, namely that *consciousness is not limited to the head*. Only the option that consciousness is present there where the attention is – in this case in the hand – can explain this difference without having to resort to rather far-fetched projections in time and space: an embodied consciousness, as we can observe already in the first brainless stages of animal evolution. This embodied consciousness is also able to expand outward, witness blind people who can feel through the tip of their cane, tennis players who feel and play the ball through their racket, the surgeon who operates on a patient through a Da Vinci robot and needs not even be in the same room, and the archer who has his consciousness out there in the bull's eye. In brief, I suggest that

6. WHERE DOES CONSCIOUSNESS RESIDE?

consciousness is not localised in the brain, because it is not a material thing, but a function, a process, a facility of the entire organism.

For that matter, Libet did view his findings as a definitive denial of the theory that brain and consciousness are identical, which is considered heresy in his specialty. For that reason, many objections were voiced by adherents to the identity theory, such as neurophilosopher Patricia Churchland,[4] who compares the findings with visual illusions. We will not go into the discussion with Libet, but will only note that he stuck to his findings and their consequences even in 2006:

> Conscious mind can only be regarded as a subjective experience, which is accessible only to the individual who has it. Thus, it can only be studied by reports given by the subject her/himself. It cannot be observed or studied by an external observer with any type of physical device. *In this sense, subjective experience (the conscious mind) appears to be a non-physical phenomenon.* (Emphasis added by AB.)[5]

Consciousness, therefore, cannot be equated with brain processes; it is something different, possibly even something that does not fit in the natural scientific model. All sensory information arrives along a variety of different paths which all require their own time span. The difference may be as much as half a second. That is a real problem, because when stimuli of two neurons do not arrive at precisely the same time at the synapse with a subsequent neuron, the latter will not fire. How do those successive stimuli all come together into one whole? This is the question of the so called 'binding problem.' There is no place in the brain where that is resolved. With his projection idea, and without mentioning the binding problem, Libet appears to have found a solution to this: everything simply adapts to the desired timing, forward or backward, it does not matter. But the binding problem exists only if we conceive of consciousness as a product of the brain. If we decide that information is not combined by the brain but by consciousness, the problem disappears.

Locally Connected

Libet's research is not unique in showing that there is consciousness in the body, which is only in the second instance 'represented' in the brain. In the 1970's – when scientists still tormented monkeys without any qualms in the name of augmenting scientific knowledge – researchers did the following experiment: they cut the bundle of sensory nerves of the hands of a number of monkeys. This bundle consists of nerves, each of which originates in a different finger, and ends in the brain, each in a specific spot in the sensory cortex. Every finger has its own spot there, neatly arranged in the same sequence as the fingers on the hand.

The nerve bundle, which was reconnected as a whole, would after a considerable time grow together again, but unfortunately the individual nerves do not look for their own corresponding extensions; they grow together arbitrarily. This therefore confuses the order of the 'wiring.' If, for instance, the feeling of the thumb were located in the brain, the result of this disorder would be that when the thumb was touched this would be felt as if, for example, the index finger were touched and when indeed the stimulus would be projected back through the nerve.

Although it is of course impossible to ask a monkey what it is feeling, it is possible to map the representation in the brain cortex. This was therefore done right before the experiment, and the expectation was that this map would become completely confused after the nerve bundle had grown together again. This would have supported the idea that the touch to a finger is 'felt' in the brain. After seven months, the sensory cortex of the monkeys was mapped anew and, to the astonishment of the researchers, touching the thumb proved to provoke activity in the 'old' thumb area; this was also true for the other fingers.[6] Although the wiring had been disordered, the connection to the locations in the brain had adapted themselves to the locations in the hand where perception was taking place. But does this also mean that the monkeys actually felt the touch in the normal place? That is the kind of information that can only be obtained from a human being.

6. WHERE DOES CONSCIOUSNESS RESIDE?

A similar experiment was performed by neurophysician William Rivers on his fellow neurologist Henry Head. They published an article on it in the journal *Brain*.[7] In 1903, Rivers cut two bundles of skin nerves of Head's fore-arm and reconnected them. Initially, there was an area that was insensible to touch and temperature, unless it went beyond a certain limit. Then the pain was excruciating. (Think of how those monkeys must have suffered!) After five months normal feeling returned to some extent, but it was hard to localise it. Ice on the beginning of the fore-arm was felt as cold in the thumb. It took 576 days before all feeling had returned and was felt in the right places. Touch to the thumb was then indeed felt in the thumb.

In both of these experiments the place of the fingers on the hand turned out to be decisive, not the connection with the brain. The latter had adapted itself to the actual situation! *What counts is the feeling in the finger, not the process taking place in the head.*

Amputations

All right, this seems clear, but aren't we overlooking a detail here? If the relative locality in the brain is restored, doesn't that mean that the locality in the brain is of decisive importance? And doesn't this then support the view that the feeling of pain is produced in the brain? An argument supporting this also seems to be the existence of chronic pain syndrome in which, after the relative injury has healed, the pain does not stop. This is caused by the fact that the brain no longer adapts itself to the healed situation. Doesn't this signify that Mosley is right when he says that his pain is located between his ears? Not necessarily. In this case too, could consciousness perhaps be the cause of the situation in the brain? It is not always obvious what is cause and effect. We often unjustifiably assume that the material situation is the cause of an immaterial result. For example, we recognise that someone is elderly by the person's wrinkles; but that does not mean that wrinkles cause old age.

Take the strongest example of chronic pain syndrome, phantom pain, when the physical location of the pain does not even exist anymore. Since this location therefore cannot be the cause of the pain, it has to be the brain that is the cause; this is what was thought in the 1990's. The theory was that the part of the sensory cortex corresponding to the amputated body part had become unemployed. This part shows spontaneous activity which is felt as phantom pain. Or in a minor variation of this theory, the function is taken over by an adjacent area.

For instance, the area that processes what is felt on the forehead lies right next to that corresponding to the thumb. When someone was touched on the forehead this would then feel as if the phantom thumb was touched. Strangely enough, in practice there seemed to be a particular correlation between touching the lip and feeling of the phantom fore-arm.[8] This is called the theory of maladaptive plasticity.

Neurologist V.S. Ramachandran has devised a brilliant therapy for phantom pain. Does he change anything in the brain? No, he tricks consciousness. In a case of phantom pain in an amputated arm, which creates the feeling as if that arm is having a painful cramp, he has his patient put his healthy arm into a kind of drawer that has a mirror on the side of the missing arm. This makes it look to the patient as if he can see the missing arm again, which is of course an illusion. By his moving the healthy arm and hand it looks as if the missing arm and hand are making the movements, and the pain of the cramp disappears.[9] The cure, therefore, does indeed go via the consciousness of the localisation, even though the latter is physically absent!

Figure 7.   Therapy for phantom pain

Ramachandran also believed in the theory of maladaptive plasticity which in the meantime, however, has been proven wrong. Researchers in Oxford have found that the representation of the arm simply remains present and grows with the level of pain that is felt. They conclude that 'contrary to the maladaptive model, cortical plasticity associated with phantom pain is driven by powerful and long-lasting subjective sensory experience.'[10] That is quite a statement. The 'powerful and long-lasting subjective sensory experience' in a no longer existing arm feeds the brain function, and not the other way around. When we combine this finding with the better known one in which it is shown that the pain and the way the position of the limb is experienced are related to the situation before the amputation,[11] it begins to look very similar to the situation in chronic pain syndrome. This would mean that it is the pain that is experienced which irritates the brain, not the reverse! But that does not take away the fact that the pain experience and the brain are in a vicious circle with each other, and that the pain disappears when we pull the brain out of this circle with pain medication. And of course, the peripheral nerve tissue in the body may also be damaged and cause pain, such as in case of a herniated or slipped disc. In that case we feel the pain along the entire nerve which, after all, forms part of the damaged cell.

Consciousness is Sometimes too Quick for the Brain

In the case of Rivers and Head, the pain evidently made its way to brain locations that had to be formed anew, just as Pedro Bach-y-Rita had to do after his stroke. It has been shown that when someone, whose movements have been impaired by a stroke, is able to make an image of a movement, he can make that movement already without laborious practice. Neurology professor Theo Mulder described in an interview[12] a patient who could hardly walk after a stroke. His leg lurched and his arm stuck out like a wing. The man had been a passionate ice-skater until his seventieth year. The physiotherapist had the inspiration to ask him, in lieu of walking, to skate on the smooth floor in his socks. It was a remarkable success.

Another patient, a former miner, had lost the sense of balance in his torso after a stroke; he was completely unable to move his arms in relation to his torso. The resourceful physiotherapist set up a kind of corridor with tables and got a little cart. The patient lay down on the cart on his back and managed, just like formerly in the mine, to do things with his arms he was completely unable to do before. Finally, there was a music director who, also after a stroke, could not move his arms. But what did he do when he said: 'I used to be able to do this'? He made typical conductor's movements with his arms.[13]

We can even make movements in which our brain has no involvement at all. Nerve cells can fire once every 8 to 10 milliseconds. Experiments have been made in which test subjects had to react quick as lightning to a stimulus. They had to pronounce letters while their lower jaw was pulled askew very quickly and briefly. The test subjects corrected this inconvenience with their lower lip, so they could pronounce the letter correctly. They did this within a time that really did not allow for the possibility of the brain to intervene: between 5 and 10 milliseconds.[14] When we speak, we use some seventy different muscles; all of these had to adapt to the new position of the mouth. In this reaction there was definitely no time available for the brain to 'figure out' how to adapt.

In their article *Ultrafast Cognition* researchers Sebastian Wallot and Guy van Orden collected many of such reaction-speed experiments in which a choice between two kinds of reactions to a stimulus had to be made as quickly as possible. The times that were observed left no possibility at all for brain processes that would have had to take place between the perception and the choice for a specific reaction. Explanations which, for instance, maintain that the nerve impulse itself already carries enough information to determine a reaction, proved to be inadequate. They concluded, therefore, that the metaphor of the brain as the central, controlling computer of consciousness is untenable.[15]

They proposed another metaphor: consciousness involves *a synergy of body, environment and mind, which therefore enter into interaction together*. Wallot and Van Orden postulate that our reactions are, as it were, embodied ahead of time and rooted in the environment, and that for this reason they can be so fast that they exceed the inadequate speed of the neurons in the brain. According to this theory, consciousness is not (solely) located in the head, but also in the body and the environment, and is not limited by the demands which neural processes have to satisfy. Let me reiterate clearly: *consciousness is not produced by the brain, but it interacts with the brain.* Together they form a synergy, as it is called by Wallot and Van Orden.

Thus it seems that the body can move without involvement of the brain; as if, again, consciousness is a function of the entire body. When a concert pianist has practised a piece often enough, he has it in his fingers, as we call it. He has to forget all his practice and let his fingers do the work. Perhaps the expression 'to have it in one's fingers' needs to be taken much more literally than we think.

Empty-Headed yet Adequate

Hence we can experience pain and make movements without a connection with functioning corresponding brain locations. It has even been shown that we can make quite radical changes to the brain without causing much impact on consciousness. Doubts about the importance of the different parts of the visual cortex arose already in the 1950's. Psychologist K.S. Lashley taught rats the way through a maze in the 1940's. He then removed part of the cortex and let them go through the maze again. The greater the part of the brain cortex that had been removed, the more difficult it was for them to find the way but, as it turned out, there was no relationship at all with any particular part that had been removed.[16] Even damage to 80% of the visual cortex did not create major visual handicaps for the rats; in the case of cats, 98% of the optical channels could be cut without serious consequences for perception. Even the combination of both interventions had little effect on visual recognition.[17]

Of course, none of this tells us much about human visual consciousness. For in animals we do not know to what extent we may have to do with a larger role of 'blind vision,' a form of perception without consciousness. Apart from the two channels from V1 to temporal, the 'what' channel, and to parietal, the 'where' channel, there is another channel that runs from the eyes to the parietal cortex, the so-called old channel, meaning older in evolution. This channel bypasses V1 and goes directly to the parietal cortex. The old channel does not generate a (conscious) image but is responsible for the phenomenon of 'blind vision.' If for some reason the occipital cortex is not functioning humans can no longer see anything, but they can still precisely point to where an object is located when a researcher asks them. The patients themselves think they are just guessing, but they always guess right. The parietal cortex is the area that is concerned with place in space. The old channel probably generates the reflex that enables us to avoid a fast flying stone that we see coming out of the corner of our eye, without having observed it consciously; and also enables us to return a fast approaching tennis ball. But why does it become conscious in the 'where-channel' and not in the old channel (the one with blind vision)? There is no observable difference in the construction and localisation of the cortex. In both cases the thalamus is involved which has been allocated a crucial role in consciousness.[18] There is no difference between neural activity that would lead to consciousness and the great majority of activity that does not. *Evidently, activity in the cortex does not necessarily mean that there is or arises consciousness.*

But it does lead to doubt as to the significance of different cortex locations. For the same is true in movement. In 1876, Professor Otto Soltmann removed the motor cortex of young dogs and rabbits, the cortex that is 'responsible' for movement; he discovered that the animals continued to be able to move.[19]

Wrong Connection

We can even add a comparable riddle to this. Why does the visual cortex produce a visual experience, and the auditory cortex one of hearing? What makes the 'visual' neurons so special? Aren't all neurons in principle the same?

6. WHERE DOES CONSCIOUSNESS RESIDE?

There has been a remarkable experiment with ferrets.[20] In newborn ferrets the visual track was severed from the visual cortex after the thalamus and connected to the auditory cortex, which is normally used for hearing. Because ferrets are very immature when they are born, they are particularly suitable for such an experiment. The ferrets could then see normally, but they used the original auditory cortex for it. This demonstrates that the visual experience is not at all dependent on the visual cortex. Neuroscientist Alva Noë concluded: 'There is nothing at all special about the cells in the so-called visual cortex that makes them visual.'[21] Neuroscientist Vernon Mountcastle found in the 1950's that the visual, auditory and sensory cortices all had the same columnar structure of six layers of neurons.[22] And all our senses translate their very different types of information into the same neural language: electrical discharges.

The conclusion appears to be that what is perceived is determined by the sense organs, and not by the regions in the brain that seem to be appointed for those functions. But even that is not true. For instance, to 'see' you don't need eyes. When you feel an object with your hands, you become aware of touch impressions. But something else also takes place: you form an image of what you have in your hands. And an image, in turn, mobilises the visual cortex. Thus you 'see' with your hands. Deaf people use the language areas for their hand gestures, among others, the centres of Broca and Wernicke, and not the areas that only govern arm movements.[23] But that this is also possible for any arbitrary area of the skin had not occurred to anyone – the precursor of the 'lollypop' of Paul Bach-y-Rita in Chapter 2 was an instrument with many little points in the back of a chair, that could prick an 'image' into one's back, and it worked. When one is looking for another sense organ, the skin is a good choice. It is our largest sense organ. And the striking aspect was that the subjects of the Bach-y-Rita experiment no longer experienced the two-dimensional feeling of the skin of their backs, but the three dimensions of the world around them. That can no longer be explained as part of evolution such as, for instance, the 'seeing' of touching fingers, since fingers anyway always mediate the perception of a form. Three-dimensional seeing takes place in our consciousness, but it is in this case not easy to explain neurologically, as a 'question of wiring.'

The Homunculus

I hope it has become clear that I do not assert that the brain does not play a crucial role in perception. We may surmise this role because of the finding described above that when pain is caused in sleep, the sensory centres are indeed activated, but not the centres that play a role in the avoidance behaviour aroused by the pain.

And if it is true that the brain does not feel anything, we should also wonder whether it ever thinks or calculates anything. Is it able to be happy or sad? Can the brain like sailing or jazz? Of course, when doing such things we do need the brain. Still, we can often read: 'the brain thinks that…' or 'for the brain it is as if…' But the brain is never hungry or angry, it never feels like running and can never perceive anything. Only organisms do that. But it takes real effort to find scientific literature in which the authors do not commit this fallacy and, according to the philosophers Bennett and Hacker, this is the principal obstacle in the way of a healthy manner of thinking about the relationship between consciousness and the brain.[24] How did this picture of the acting and thinking brain ever come into the world?

For instance, Paul Bach-y-Rita said that we see with our brain, not with our eyes. He meant that it makes no difference how a visual stimulus comes in, whether it is via the skin, tongue surface or retina – all of them two-dimensional! – everything ultimately ends up in the visual cortex. However, we have just seen that it doesn't even matter where the stimulus ends up in the brain. There is nothing visual about the neurons in the visual cortex.

The fact that the two-dimensional picture is transformed into a three-dimensional one is ascribed to the brain. And this high-handed intervention by the brain into the picture that was observed is supposed to have a side effect, namely of visual illusions. But in Chapter 1 we have seen that visual illusions have much to do with

6. WHERE DOES CONSCIOUSNESS RESIDE?

culture, and are thus perhaps more a consequence of the learning process of a precocious consciousness than a precocious brain. These visual illusions are one of the reasons underlying the dominant idea that our brain forms an unreliable picture of reality. The whole path from the eyes to the visual cortex with associated areas leads us to presume that the light that falls on the retina only comes together as a picture in the brain. The brain would then see to it on the way along this path that out of insufficient information a complete picture comes into being. The brain is supposed to do this like a kind of biological computer. The idea that a picture arises in the brain was created by Descartes who, thanks to the invention of the camera obscura, imagined it as shown below.

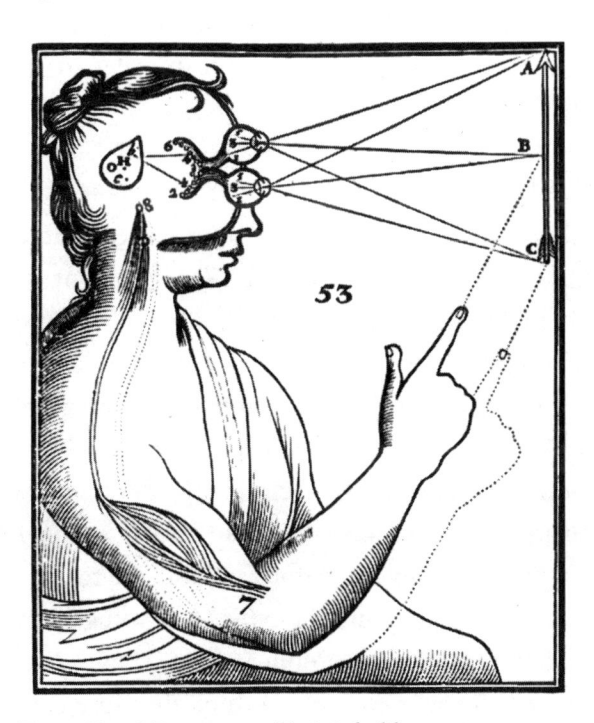

Figure 8. From René Descartes, *Traité de l'homme*

The leaf-like structure in the brain represents the pineal gland where, according to Descartes, the spirit makes contact with the brain and sees the image that comes in through the eyes. Thereupon the spirit

75

puts the body to work, via the pineal gland, in order to perform an action. Neuroscientists call this body-spirit dualism, and Descartes' greatest error. Daniel Dennett thinks dualism is a 'lost cause' that 'has to be avoided at all cost.' And Antonio Damasio even called one of his books *Descartes' Error*. The spirit in the pineal gland plays the role of spectator in the brain, and in neurophilosophy that function is often called the homunculus.

This problem was created by educational pictures which often showed the path followed by an image ending in the back of the head, where a spectator watches the image on a screen. Consciousness is thus pictured as an imaginary figure that watches images made in the brain. But there is no homunculus, no figure in the brain, say the justified opponents of this idea, because this homunculus would have to have another homunculus in his brain in order to see anything, and so forth.

Therefore I Am My Brain

But how does this all work? The solution embraced by most neuroscientists is very simple: the pineal gland of Descartes is expanded to include the brain as a whole. The homunculus now fills the entire skull. It is the brain that sees an image, hears, feels, knows, remembers, wants, rejoices, wants to go on vacation or see a movie, and likes avocado. *We are our brain*. It is a new dualism, the brain-body dualism. Ultimately, Descartes is also responsible for the idea that we are our brain due to his best-known words: *Dubito ergo cogito ergo sum* – 'I doubt, therefore I think, therefore I am.' The premise is that I am merely my thoughts. According to conventional neuroscience, my thoughts are produced by my brain. Hence we may now say: *Cogito ergo cerebrum sum* – 'I think therefore I am my brain.'

Descartes once introduced the picture of images in the brain, and it seems that we are having a hard time getting away from it. Even if we don't accept that reality is the product of the brain, according to this notion the brain still has to be filled with representations of the world and of our body. Accordingly, the projection in the visual

6. WHERE DOES CONSCIOUSNESS RESIDE?

cortex in the back of the brain (V1), which can be made visible with the VEP test (see Chapter 4), is described as an image. The fact that the associated areas, which are later touched on by the visual path, return branches to V1 could create the impression that an image arises there, against the back of the head. And indeed, the researcher who looks at the results of the VEP on his screen, witnesses something there that could be interpreted as the electrical construct of an image.

The question is then whether the brain also sees that image. An image cannot see itself, can it? When we assert that the brain sees the image, or when we talk about representations in the brain, in fact we revive the homunculus, who would have to have his own homunculus, and so forth, as we saw above. Neither does psychologist Kevin O'Regan – a confirmed materialist according to biologist Rupert Sheldrake who quotes him – accept that an image arises in the brain, for that would 'put you in the terrible situation of having to postulate some magical mechanism that endows the visual cortex with sight, and the auditory cortex with hearing.'[25] Where then does the image appear? To be simple and practical about it, shouldn't we say: in our consciousness? If a neuron can't see or hear, what is its role?

The Near-Death Experience of a Nerve Cell

What is it actually that distinguishes a nerve cell from other cells? A nerve cell conducts an electrical stimulus. That is why we can make an EEG (electro-encephalogram) of the electrical activity of the brain. There is also another organ the cells of which conduct electrical stimuli, namely the heart. The heart cells function, as it were, also like nerve cells. But how does this conductivity work? Every animal cell maintains a difference in potential on both sides of its membrane as long as it is alive, by keeping out charged sodium and calcium atoms (ions) and keeping in potassium ions. When it dies, this difference ceases to exist. The cell *depolarises*, and the charged atoms that were at first kept out, such as sodium ions and calcium ions, flow into the cell. Well, that is precisely what happens in a nerve cell when it is stimulated, and the depolarisation travels along the entire nerve fibre like a depolarisation wave. This is the stimulus that is conducted. But the nerve does not die; behind this wave the difference in potential is restored as quickly as possible. That takes a little time and much energy. One could say that the nerve cell *temporarily* and *locally* dies. The nervous system is constantly balancing on the edge of death. The nerves can even actually die when they are overstimulated. This is called exitotoxicity, and to prevent it extra nerves are engaged at the synapses that suppress the stimulus by means of the neurotransmitter GABA (gamma-amino-butyric-acid). These nerves are stimulated by medications like tranquillisers (valium) and sleep aids.

So the image appears undeniably in our consciousness but, equally undeniably, after it has been processed in the brain. Does this mean that consciousness resides in the head after all? Let's take a look at that. It is very helpful to follow the path traveled by the visual stimulus from the retina in the eyes through the brain into the visual cortex in the neuroscientific manner. This description can be understood as if it were a physical phenomenon like the light in a camera. In this picture, only consciousness remains a riddle. Would that were the only problem, however. This description of the path from the eyes to the visual cortex is but half of the story. Seeing not only requires much more than a functioning physical organ and visual cortex, as the case of the blind and seeing woman in Munich has taught us; it also involves another aspect which we have so far neglected to mention, namely attention.

Attention

What we perceive depends in great measure on our attention. Attention may be viewed as directed consciousness. We direct our consciousness to something, and this direction is called attention. The direction may also fade, with the result that attention disappears and what is left is undirected, daydreaming consciousness. With our so-called conscious senses we primarily perceive that which has our attention, which we want to perceive. And the rest then no longer penetrates into our consciousness. (There exists a world record of uninterrupted television watching. The person who achieved the

6. WHERE DOES CONSCIOUSNESS RESIDE?

latest record of 86 hours, Efraim van Oeveren, said that after 48 hours he no longer registered anything of what he saw.)

Fifteen years ago this was demonstrated by the well-known experiment with the invisible gorilla. Test persons were shown a short movie in which two teams of three basketball players each, dressed in white and black, are playing with two balls for a few minutes. The test subjects had to count the number of passes. In the course of the game a character dressed up as a gorilla walks around for a while, conspicuously drumming his fists on his breast; then he disappears again. After the movie the test subjects were asked two questions: how many passes did you count? and: did you see the gorilla? I once saw the movie in an auditorium filled with physicians. Virtually the entire hall missed the gorilla. I had read something about it, so I didn't get caught.[26]

We have all kinds of different shades of attention; we can roughly distinguish two kinds: attention for details, and attention for totalities. The first kind, focused attention, is inevitably, almost lawfully, accompanied by a certain blindness for new information,[27] as demonstrated by the invisible gorilla. How this works and why one door is open and the other closed will be discussed in Chapter 12.

Not everyone sees the same things, but that may of course be caused by differences in brain connections. However, seeing may be trained, which is something we do ourselves with the aid of our consciousness. We are even able, without making any changes in the eyes, to improve our sight by a change in mindset, by a different psychological attitude. The smallest print at the bottom of the card the ophthalmologist gives us to read may be too small for us, but it can be read without effort if, unconsciously to us, someone suggests that we have acute vision. Also, if the same small print is placed at the top of the card, it suddenly becomes readable.[28]

Where then Should We Look for Our Consciousness?

Hence, attention, or intention, determines whether we see something. But where do we see a thing, or the blue sky? Where do we hear the rustling of tree leaves? Where do we feel pain in the skin of our calf?

Do we see, hear and feel those things between our ears? Do we actually feel the pain of our calf in our brain? Do we first take the blue sky into ourselves in order to perceive it in the back of our head? Or do our eyes offer our consciousness access to the blue sky there where we see it, and similarly to the tree there where we hear the rustle of its leaves? What was the moment when the brain began to manufacture an unreliable picture of reality? What could be the evolutionary advantage of that? In one way or another there has to be some kind of two-way traffic in seeing and other forms of perceiving, if we say that consciousness arises in the brain. First the inward path that can be scientifically described. But why do we still see the tree outside? According to Benjamin Libet, mentioned above, this is called outward projection. This outward projection would then be a second movement, upstream in the causal sequence, namely the movement that enables us to experience the tree *where it stands*, and pain *where it is caused*. This movement cannot be scientifically described.

Maybe it is now possible to give the following idea a chance. Since we have no notion whether, and how, the brain produces consciousness, and since the supposed physical nature and location of consciousness are unsolved riddles ('what does consciousness consist of?'), isn't it much simpler to situate consciousness with just as much justification there where we experience it, namely in the world – the blue sky, the rustling leaves – and in our body – the pain in our calf – and, in the acts of thinking, reflecting and imagining, in our inner world? But even the word 'situate' is not correct here. Consciousness is not something in a location; it is activity. When we are in Chicago and think of London, we don't need to go to Westminster Abbey to find our consciousness.

That is not naïve; we find this idea in different words in the work of philosopher and psychologist William James (1842–1910) who said that the idea that an image is seen in the brain 'violates the reader's sense of life, which knows no intervening mental image but seems to see the room and the book immediately just as they physically exist.'[29]

6. WHERE DOES CONSCIOUSNESS RESIDE?

Philosopher Alfred North Whitehead (1861–1947) maintained that projection plays a role in perception: 'Sensations are projected by the mind so as to clothe appropriate bodies in external nature.'[30]

Philosopher Henri Bergson (1859–1941) had the following view: 'The truth is that the point P, the rays which it emits, the retina and the nervous elements affected, form a single whole; that the luminous point P is a part of this whole; and that it is really in P, and not elsewhere, that the image of P is formed and perceived.'[31]

Bergson's colleague Maurice Merleau-Ponty (1908–1961) just says: 'We should therefore not wonder whether we really perceive a world; on the contrary, we should say: the world is that which we perceive.'[32]

Cognitive philosopher Alva Noë (1964) also has the opinion that consciousness does not reside in the head: 'The world is not a construction of the brain, nor is it a product of our own conscious efforts. It is there for us; we are here in it. The conscious mind is not inside us; it is, it would be better to say, a kind of active attunement to the world, an achieved integration. It is the world itself, all around, that fixes the nature of conscious experience.'[33]

Psychiatry Professor Thomas Fuchs is perfectly definite about it: 'The brain is certainly a central organ of the living being, but it is only an organ of the mind, not its seat. For the mind is not located in any one place at all; rather, it is an activity of the living being which integrates at any moment the ongoing relations between brain, body and environment. Assuming such an embodied, extended and dynamic view of the mind, the brain loses its mythological powers and turns into a still fascinating, yet far more modest mediator of human experience, action and interaction.'[34]

Consciousness, therefore, does not reside in the head, at least not when we perceive something. But when we imagine or visualise something, say with our eyes closed, or when we think, don't we then speak of an inner experience? Where then do we find this 'inner?' When we close our eyes and imagine a mountain landscape, where do we see that? On the inside of our forehead or eyelids? As we will notice, it does not fit there. We see it 'outside,' there where we would see it if it were reality.

81

Does consciousness exist in any one place? It may be difficult to let go of the notion that we experience light and sound in our head. After all, our eyes and ears are openings in our skull. Except for the cornea, embryonically the eyes are even part of the brain. The location of consciousness outside the head becomes obvious when, with our eyes closed, we try to determine the form of something by touching it with our fingers. This is felt distinctly outside the head, at the spot where the tangible form is.

Without any doubt, our brain and sense organs play a crucial role in our consciousness of the world and our body, and the importance of developing an understanding of this cannot be overestimated. But based on the conviction that maybe we will someday know how the brain produces consciousness, is it right to set up a premise that consciousness is enclosed in our skull, and that we are our brain? Wouldn't it be better to speak of sense organs and brain as, respectively, portals and instrument of consciousness? That generates a picture which right away indicates dynamics, activity. Consciousness is not something static that is simply there. Consciousness is something we do, *an interaction enacted between the world, our body and our inner world*. But then consciousness is not in any fixed location at all, for it is not a thing. It has no quantity, only quality, namely the content of consciousness. There is no consciousness without content. Even when individuals experience an empty consciousness, for instance, in meditation, this emptiness is the content of their consciousness.

We Are Embodied Beings
Consciousness, therefore, is not located in our skull but can, in a manner of speaking, reach from our body to the stars. For it is not limited to any one location. The notion that consciousness resides in the head is behind the fact that most brain books pay no attention to the rest of the body. The idea that the brain has no need of the body is even the subject of an imaginary neurophilosophical experiment with the object of demonstrating that the brain is a computer which represents reality.

CHAPTER 7

The Brain in a Vat

The only purpose of the body is to feed, transport and propagate the brain.

Dick Swaab[1]

When I was in high school in the 1960's I was an avid reader of the disturbing short stories Roald Dahl wrote before he limited himself to children's books. One of these stories I have never forgotten is *William and Mary* from the collection *Kiss Kiss*. It is about a recent widow who received a letter from her lawyer, written by her deceased husband before his death, in which he told her that he had given instructions to have his brain removed by an ambitious experimenting physician immediately before his death. The brain was to be kept alive with the aid of a kind of heart-lung machine. In this way the brain should be preserved 'alive and conscious' in a tank in the laboratory of this 'mad scientist,' and in the letter he asked her to come and visit him. Actually, she had been much relieved to be rid of him – he had badly dominated her and forbidden her to do a number of things, including smoking cigarettes – so this was something she had to think about. In the end, she decided to do it, for she no longer had anything to fear from him. She visited the laboratory and discovered that the brain still possessed one eye. The doctor showed it the newspaper, *The Mirror*.

'He hates *The Mirror*, give him rather *The Times*.'

'Very good, Mrs. Pearl, we will let it read *The Times*. Of course we want to do all we can to make it happy.'

'*Him*,' she said, 'not *it*. *Him*.'

When leaving, Mary looked once more into the eye while she brought a burning cigarette to her lips ... 'and right away there was a flash in William's eye ... The pupil contracted into a tiny pin's head of terrible rage ... She inhaled deeply ... and then suddenly, woosh, she blew the smoke through her nostrils – two narrow plumes of smoke that rippled the water, spread like a thick, blue cloud over the water and hid William's eye from view.'

A Neurophilosophical Vat

There circulates among neurophilosophers an imaginary experiment of a 'brain in a vat,' a brain floating in a liquid and receiving 'sensory' information through a computer, so that much more comes into it than through that one eye in Roald Dahl's story. The idea is attributed to Daniel Dennett who began his book *Consciousness Explained* with this picture.[2] Others ascribe it to philosopher Hilary Putnam who indeed published this imaginary experiment much earlier:

> ...imagine that a human being (you can imagine this to be yourself) has been subjected to an operation by an evil scientist. The person's brain (your brain) has been removed from the body and placed in a vat of nutrients which keeps the brain alive. The nerve endings have been connected to a super-scientific computer which causes the person whose brain it is to have the illusion that everything is perfectly normal. There seem to be people, objects, the sky, etc; but really all the person (you) is experiencing is the result of electronic impulses travelling from the computer to the nerve endings. The computer is so clever that if the person tries to raise his hand, the feedback from the computer will cause him to 'see' and 'feel' the hand being raised. Moreover, by varying the program, the evil scientist can cause the victim to 'experience' (or hallucinate) any situation

or environment the evil scientist wishes. He can also obliterate the memory of the brain operation, so that the victim will seem to himself to have always been in this environment. It can even seem to the victim that he is sitting and reading these very words.[3]

However, for Putnam this imaginary experiment led to the conclusion that the idea of the brain and consciousness being identical is invalid, in contradistinction to Dennett's contention. The premise of Roald Dahl's story is also that the brain suffices to represent a 'person.' It is interesting that it then evidently also suffices that the brain only receives information. These stories do not speak of any output, perhaps with the exception of William's pupil. For that matter, neither Putnam nor Dahl are mentioned by Dennett.

This imaginary experiment was the inspiration for the movie *The Matrix*, which was about the very same problem, and resulted in a fascinating spectacle, and also a neurophilosophical headache. For the problem that emerges there is the question of how, if we can fool the brain in this way, we can be certain that we are actually in contact with reality. Doesn't our brain make, or at least recreate, reality? Doesn't it all take place in the head? Isn't it all a bunch of illusions that we experience as reality? Aren't we also in a Matrix situation? The general view is that we will never know this. But we can only entertain this disturbing thought if we believe that consciousness resides in the head.

When someone writes a book with the title *We Are Our Brain* (Dick Swaab), we could anticipate that in his view of the human being the body comes off badly, as shown by the quotation at the top of this chapter. It comes pretty close to a model of a bodiless stand-alone brain.

Someone once asked Dick Swaab whether it would be possible for the brain to stay alive in a vat while maintaining consciousness, assuming every necessary demand would be met, such as nourishment and the supply of information from a computer that simulates the outside world. For wouldn't this be the logical consequence of his point of view? 'I consider that theoretically quite possible,' was his answer, 'but why would you do it?' The person who asked the question had justifiably drawn the conclusion that for neurodeterminists we are in fact just a brain in a vat, the vat being the skull and the body the machinery that has to keep the brain alive.

Boss in Our Own Brain

Soon after this conversation, something remarkable happened. Dick Swaab was interviewed on television in 2010, and I was surprised to hear him say: 'I do prefer to be the boss in my own brain.' I couldn't believe my ears. Did he really say that? I would never have expected that of him. For then he actually considers his *I* and his *brain* as two different things. Or should I perhaps conclude that for him too, in daily expressions of common sense, *brain* and *I* are not the same thing, but only in his beloved theory that we are our brain?

Actually, the problem of the brain books is not that the facts are not stated right, but that facts always have to be interpreted. Nietzsche even said: 'There are no facts, only interpretations'. By the way, that also goes for the book you are now reading. But in a materialistic book, preference must of course be given to a scientific, meaning materialistic, theory over any possible demands of common sense. That is not so strange, because many results of scientific research are counter-intuitive. And the notion of common sense, namely that we ourselves determine what we want to think, is something with which neurodeterminists cannot agree. Be that as it may, in normal daily life Dick Swaab turns out to be someone who also uses his common sense and therefore wants to be boss in his own brain.

This is characteristic of the position of neurodeterministic authors of brain books. What is asserted often does not apply to the author himself. The journal *Nature* of June 2011 had a great example. Philosopher Anthony Grayling wrote there that our brain produces religious faith and superstition because at some point this had become hardwired into it by evolution:

As a 'belief engine', the brain is always seeking to find meaning in the information that pours into it. Once it has constructed a belief, it rationalises it with explanations, almost always after the event. The brain thus becomes invested in the beliefs, and reinforces them by looking for supporting evidence while blinding itself to anything contrary.[4]

Of course, this in no way applies to Grayling's own brain.

The Head in a Vat already Exists
In the United States there are already more than two hundred people who had their bodies frozen through cryonics after death. Before their decease they were hoping that in the future a technology would be developed by which they, after their bodies were thawed, could come to life again. For this reason, the organisations that do this, Alcor Life Extension Foundation and Cryonics Institute, don't call them 'deceased' but 'patients.' It goes without saying that it costs a lot of money. People with a more modest budget are offered a less expensive option; they can have their head frozen and preserved in the hope that, once we have made some progress again, a solution may also be found for it, such as connections with a computer or a 'body transplant.' But what if their hope is indeed fulfilled? Can they have a human consciousness without a body? And are they without a body still the same person? What role does the body play anyway?

The condition of the thawed head – assuming it were really possible – would somewhat resemble the following. In Lausanne, Switzerland, people are working on a project to build a supercomputer in about ten years, that will equal the neural network of human brains, cell by cell and synapse by synapse. It is called the Human Brain Project. It is expected that this computer will produce human consciousness. But 'human consciousness' is an abstraction; in practice the only thing that exists is your and my human consciousness.

And that is individually embodied consciousness. This brain computer will never have reason to be hungry or tired, suffer pain, be happy, to lie, in brief, to show anything resembling a human condition, because it will never be able to do anything with a body, certainly not the most free things that make a human being human, such as playing, art or sports. It will never show such human reactions as laughing, crying, blushing or becoming depressed. Besides, how would neurotransmitters be built into that computer?

The same holds true for the thawed head. It would not be able to feel anything; not only that, it would most probably not understand anything and it would certainly have no will. And if the thawed head – suppose it ever came to that – were given another body, it would not recognise itself, not merely as a body, but neither as a person. Why not?

CHAPTER 8

Consciousness Needs the Body

The spirit does not live in the body, but the body lives in the spirit.

Hildegard von Bingen

Neurologist Antonio Damasio[1] defends the point of view that the way we feel, such as the mood in which we wake up, is determined by the physiological condition of our body. The autonomous nervous system plays an important role here. This physical feeling at the same time generates the feeling of being embodied, which may well be the principal support of our self-awareness – the feeling that it is 'I' who is feeling this, thinking this and acting that. What's more, in the footsteps of the famous psychologist William James, who lived around the turn of the twentieth century, it is Damasio's opinion that what we call feelings are actually a process of becoming conscious of emotions, for instance, in the form of the thought 'I am afraid.' Emotions are physical occurrences that need not at all take place consciously, such as heart palpitations, perspiring, gasping, butterflies in the stomach. Animals also have emotions, but no feelings. Only human beings can become so conscious of these things that they recognise such physical disturbances as feelings. William James stated that emotions remain flat when the body does not contribute perceptible reactions to them:

> Common sense says, we lose our fortune, are sorry and weep; we meet a bear, are frightened and run; we are

insulted by a rival, are angry and strike. The hypothesis here to be defended says that this order of sequence is incorrect, that the one mental state is not immediately induced by the other, that the bodily manifestations must first be interposed between, and that the more rational statement is that we feel sorry because we cry, angry because we strike, afraid because we tremble, and not that we cry, strike, or tremble, because we are sorry, angry, or fearful, as the case may be. Without the bodily states following on the perception, the latter would be purely cognitive in form, pale, colourless, destitute of emotional warmth. We might then see the bear, and judge it best to run, receive the insult and deem it right to strike, but we could not actually *feel* afraid or angry.[2]

Naturally, James is not saying that losing a fortune plays no role in becoming sad. Physical reactions do not arise without reference to events; but they are needed to experience feelings.

The Soul

Emotions each have their own place in the body. How do we explain that? What exactly is the relationship — so often professed in words — between body and soul or psyche? The fact that the image on an fMRI scan of physical pain is comparable to that of emotional pain was a puzzling surprise for researchers in 2011. Pain from a heat stimulus showed virtually the same brain activity as social exclusion.[3] How can that be? Isn't the soul, of which we have an abstract and incorporeal notion today, something entirely different from the body, even if they are somehow related?

If there actually is a cohesion between the soul and the body, then the whole body is 'ensouled.' That might be the reason why physical pain and psychic pain are processed in the same manner by the brain. But then we should also have to be able to feel our soul physically.

Of course, 'soul' is an old-fashioned concept. It has no place in science. Both in older texts and in everyday colloquial speech the word is used in different ways. It might mean the same thing as what I will discuss later in this book as the 'self.' Sometimes we hear of someone who is an 'old soul,' a concept that comes from the idea of reincarnation. It may also be used as a synonym of psyche, consciousness, the realm of experiences, and it may, as a third possibility, mean a kind of bodily conglomeration of conscious and unconscious emotions that might be described as heart-and-mind, which we become aware of as residing in the breast. I am using the word here in this last meaning. Can we *physically* experience the soul in the sense of a deep-seated state of mind? Yes, we can, even in a different way than by feeling heart palpitations or changes in breathing.

It is remarkable, actually also hopeful, that research is now finally being done into physical experiences of emotions. Anger, fear, pride, surprise and other emotions are each experienced in the body in a different pattern. What gives us the feeling that our chest swells, or of a sinking heart? Can that be adequately explained by nerve networks or neurotransmitters? It has been shown that these emotions are universally experienced in the same manner in the body, at any rate in Europeans and East-Asians (see Figure A on the inside front cover). Test subjects had to read stories or watch short films that had to generate certain feelings. They were also asked to live into certain feelings. Then they had to indicate in colour in pre-drawn human figures where they experienced these feelings. The results were virtually identical for all of them. One of the most interesting aspects described by the researchers who published these body maps of emotions is that the emotions were indeed experienced in a particular part of the body, but appeared to be completely unrelated to any form of physiology. (By the way, I don't think that is true for the feeling of shame which often causes blushing.)[4] What then is acting there, if it is not something physiological or physical? The soul?

This is of course a completely different story from the neurodeterministic explanation that not our physical reactions but neurotransmitters autonomously cause our feelings, or even *are* our feelings. For instance Professor René Kahn wrote the following about the influence of dopamine (a neurotransmitter) on the nucleus accumbens (a grey matter area inside the brain):

> This experiment demonstrates that the accumbens is only engaged when we eagerly want something *and* have to make an effort to obtain it. In other words, this nucleus becomes active when we are motivated. It is of course the other way around: when this area is active, we are motivated; it is then that we really will something, that we strive, that we make our best effort ... In the accumbens dopamine is the substance that prods us to obtain what we want.[5]

Thus it is not the promised salary, the pretty car, the attractive potential partner that motivates. 'It is the other way around': the dopamine in the accumbens does this; only then do we recognise the car, the man or woman or the money as attractive. We do not love our children for who they are, it is oxytocine that prods us. We do not suffer from depression because we have lost our job, but because of a shortage of serotonin. Without any doubt, neurotransmitters play a role in those feelings. But the above representation is without any doubt equally a compulsive materialistic reversal of things. In this way a headache is the consequence of a shortage of paracetamol (Tylenol).

It is absolutely unclear what the role of neurotransmitters is in the experience of feelings. Single-cell organisms already possess these same substances. But I can't really picture the feeling life of an amoeba.

Facial Expression

The thawed head from the previous chapter, therefore, would most probably lack any feeling. It could possibly still possess the expressive muscles in its face. Our knowledge of mirror neurons tells us that we recognise the feelings of another person, at least in part, because we almost imperceptibly imitate the person's facial expression in our own expression. This would have to mean that faking an expression should generate the corresponding feeling in ourselves. As it turns out, it does indeed. As we saw in Chapter 3, test subjects who clenched a pencil between their teeth across their mouths without letting it touch their lips, found some cartoons much funnier than other subjects who had to point the pencil straight forward with their lips.[6] When you try it yourself, it becomes clear which facial expression is being imitated and which one is impeded. Botox injections in the forehead, which make frowning impossible, resulted in decreased activity in the amygdala (see upper illustration on the inside back cover.) when the participants had to pull a sad face, but not when their expression had to be a happy one.[7] We might note here that the activity of the amygdala indicates the presence of emotional reaction. In another experiment, participants who had been botoxed in the same place needed more time to catch the gist of sad or angry sentences, but not of sentences with a happy connotation.[8]

The Metaphor Machine

The body not only influences our moods, but also our thinking. Neurologist V.S. Ramachandran discovered that an injury to the left *gyrus angularis* (see Figure B on the inside front cover) resulted in incomprehension of metaphors and abstractions. And what is language without metaphors and abstractions? This is interesting because this area had been known for more than a century as the area that relates to the *body schema*, which is the name of the knowledge we have of the place and position of all parts of the body. It enables us to bring the tip of our index finger to the tip of our nose with our eyes closed, and to scratch an itch on our back. This *gyrus angularis* is also full of mirror neurons, so that we can easily imitate the positions and actions of others. Now, what have abstractions and metaphors to do with the body schema? It turns out that a great many abstractions and metaphors, with which we attempt to express and understand the relationships between objects, are derived from the body and the position of the body in space. (The *gyrus angularis* borders on the speech recognition area of Wernicke. Perhaps this has something to do with the easy relationship that is shown to exist between physical experiences and metaphorical language.)

The first researchers to point to this relationship between the body and thinking were Lakoff and Johnson.[9]

Understanding with the Body

In the meantime it will have become clear that we would never be able to develop mirror neurons without a body. In the interaction with another, the body plays a role at least as important in our understanding of the other as the brain. We can understand each other only as embodied beings. We do not need to figure out what people mean when they cry and, without having to give it any thought, we may feel the need to console such a person, for instance, by putting our arm around their shoulders. This cycle of events not only takes place in the mirror neurons, but in the entire body. The oldest form of understanding is with the body, without language, without concepts, the way animals do it.

What actually is understanding? When do we know that we understand something? Understanding really comes down to the ability to connect a feeling to a new thought. I can't find another way to say it: it is the feeling that you understand something. For sometimes a single remark, which actually has nothing to do with the subject, can give us the feeling that we understand it. And for feeling, said Damasio, we need a body. But what causes the feeling? The feeling arises when the elements of a new thought can be compared with familiar elements: 'it is like…' This is the reason why we often use metaphors in the hope to generate understanding. And from where can we better derive these metaphors than from our trusted body?

I can illustrate this with a few metaphors of different kinds.

- We use our *own movements* to clearly express things: going through something, being out of step, letting something pass,

reaching something, keeping up with something, shrinking back, catching something, following a thought.
- We do the same kind of thing with *physical positions* when we localise something: out of reach, beneath me, my standpoint, progressive, profound, the underlying concept.
- We refer to *instruments and dexterity* when we manipulate someone, handle something, hammer on something, rake stuff together.
- Finally, we use *observations* when we say: I finally see the point, I have a different point of view, a light dawned on me, a dirty business, to focus, a clear vision, a rough customer, a slippery character, a crying injustice, a disgusting sight, an oppressive atmosphere.

All of these indicate something other than the literal meaning, namely a metaphorical meaning – metaphors therefore. Undoubtedly we can all add many to the ones mentioned here. What is interesting here is that – in view of the involvement of the *gyrus angularis* –we can evidently only understand these imaginary movements, observations and positions if we have not lost the connection with our body.

Embodied Ethics

The connection between body and consciousness is even more concrete in the specialty in psychology that has developed in the slipstream of research into unconscious influences: *embodied cognition* or *embodiment*. It turns out that unconscious influences on the body have consequences not only for feelings, but also for behaviour. And it is no accident that we are all familiar with these influences as metaphors, in virtually every language. Social psychology is said to have shown that having a warm cup of coffee in our hands generates warm and generous feelings toward strangers,[10] and that a comfortably heated room gives people confidence in others.[11] People who felt excluded estimated the temperature in the research space at an average of 5°C lower than others.[12]

A hard seat causes people to negotiate harder than a soft seat.[13] Having a heavy object in our hands causes us to judge a decision we have to make more important, or an amount of money more valuable.[14] Research into the influence of the intensity of light on honesty showed that darkness leads to a lowering of our moral standards. In muted light the test subjects committed more deception than in bright light. And sunglasses also caused more antisocial behaviour.[15]

Expressions like a *clean conscience* or *pure moral standards* also prove to be metaphors that are deeply anchored in our inner life. Pontius Pilate and Lady Macbeth, who washed their hands in innocence, thus committed a universally human deed. In a space that smells of cleaning substances people act more socially and generously.[16] Test subjects who were first asked to remember unethical deeds, then had to make words of which they were given the first and last letters. They came up with words like *wash* and *soap* significantly more often than the people in the other group, who had to remember deeds of high moral standing and then produced words like *wish* and *step*. In a subsequent experiment, people who had to think back of an unethical past more often (75%) preferred a sterile piece of cloth over a pen as thank-you gift than the people who had to remember an ethical past (37.5% chose the piece of cloth).[17] People who had just washed their hands thought of themselves as more morally pure and sublime than when they had not done this; test subjects who had first washed their hands judged questionable behaviour (smoking, use of drugs, pornography, swearing, littering and adultery) more often as immoral than others, and they also considered themselves as morally better than others.[18]

We might well imagine that unethical actions give us dirty hands, as indicated by our language, and that lies or malicious words defile the mouth. They do indeed. Test subjects were told when role-playing to speak a malicious lie into a voice message or write it in an email. When they were asked to evaluate consumer products, the voice-group preferred mouthwash and the email-group products for hand hygiene.[19] It is beginning to look as if the genius of centuries-old expressions in language, to which these experiments refer, provides testimony of great insight. Our sense functions let themselves be used with the greatest ease in psychological and moral metaphors. Or should I say perhaps that our sense organs communicate more than just bare facts?

Weekly we can read reports of new research into the relationship between physical experience and behaviour. The field of research into 'embodied cognition' has become a kind of playground. But still, there is so far, in neurodeterministic terms, no indication of any theories being formed. What really is this relationship, how does it work? As long as it is assumed that the brain produces consciousness, that question is hard to answer. Only when the body is understood to be animated by a soul, or consciousness to be embodied – something that is no problem for psychologists, witness the term 'embodied cognition' – will the contours of a theory become visible.

Psychosomatics

The fact that, on the other hand, consciousness also has great influence on the body may be assumed to be common knowledge. Here also the brain clearly forms the connection between consciousness and body. Blushing, heart palpitations, erections, are immediately noticeable effects of consciousness on the body. Consciousness can also make the body ill. Anxiety and depression have a disastrous effect on the immune system. This would have to mean that a change in mindset can also have a healing influence. This proves to be true. In popular speech, a heart attack has its cause in emotions, at least until recently; these days everyone ascribes it to cholesterol. But the older view is not far wrong: anxiety and depression have been proved to cause a stiffening of the blood vessels, also in the heart.[20] And does undertaking psychotherapy make a difference? Yes indeed, cognitive behavioural therapy makes a difference in disorders of the heart and blood vessels.[21]

Gut Feelings and Love of the Heart

Another example of popular wisdom, which almost ended up in the waste basket of the natural-scientific world view, has to be saved, namely that our organs actually play a real role in consciousness. The neurotransmitters, the neuromodulators and the hormones that have an interactive function in our moods, do not occur just in the brain. They are produced in many different places in the body. Of course it has been known for a long time that adrenalin and cortisol, which are produced by the adrenal glands, play an important role in cases of stress. By the way, the adrenal glands also produce testosterone, just like the testicles. The ovaries produce estrogen and progesterone. Just like neurotransmitters, all these sexual hormones have a distinct influence on the emotional state of the body, and therefore also on mood and feelings. Thus an expression like 'he has balls' is quite adequate.

The signaling substances of the immune system (cytokines) also influence the brain. They have been shown to be able to cause depression, which is explained by the fact that they reduce the activity of the reward system.[22] And the immune substance interleukin-6 has a negative influence on the quality of sleep,[23] while it causes sleepiness during the day.[24] Interleukin-1, an immune substance which, just like interleukin-6 causes fever, does the same thing. Anyone who ever had the flu will recognise that pattern.

It is much less known that other organs also produce substances that play a role in the brain. The fact that the heart continues to be connected with love is much more than mere poetic licence. For the heart produces oxytocin, also called cuddle or attachment hormone.[25] Moreover, the heart chambers produce ANP (atrium natriuretic peptide), which has a 'calming' influence on the hypothalamus and amygdala. ANP works on the fluid balance and the sodium-potassium balance, but it also works as a brake on stress reactions such as panic disorders.[26] It is primarily produced when we exert ourselves. It was recently reported that the heartbeat also influences the level of attention.[27]

Why do we have butterflies in our stomach when have to perform, or address a seemingly unattainable loved one? Why do we have nightmares after a overly-rich meal, or diarrhea when we are nervous? These things all look like reactions of the metabolic system to signals from the brain. But the wall of the digestive tract has its own, almost autonomously functioning nervous system all the way from the esophagus to the anus. It contains 100 million neurons, as many as the spinal cord. The connection with the brain is provided by the vagus nerve. Not so long ago it was discovered that this nerve hardly gives any signals to the intestines, but that 95% of the nerve fibres send them in the other direction. In this way the intestinal track turns out to influence the brain. The genius of the English language has always known this, and tells us that the intestines are able to provide information to judge a situation: 'gut feelings.' The intestines produce a lot of serotonin, about 95% of what the body contains. This could also have an effect on our mood. Actually, all the messenger substances we know of in the brain – the neurotransmitters and neuropeptides – are also present in the intestines.

When we think of the anatomy of a worm (an intestine with a skin around it) we might well imagine that the enteric nervous system, as the nervous system of digestion is called, is older than the central nervous system, the brain and the spinal cord. It belongs to the autonomous nervous system. One of the most interesting group of substances made by the intestines is that of the benzodiazepides. We are familiar with these in the form of pills: valium and a host of other tranquillisers and sleeping pills.[28] The intestinal flora has great influence on this as Professor Geert d'Haens explains:

> Mice with a particular bacterial profile, a shortage of certain beneficent intestinal bacteria, were more aggressive and suicidal than mice with a normal intestinal flora. They generated more stress hormones. However, when normal, beneficent micro-organisms were added

to their food (such as *lactobacillus rhamnosus* which occurs in some yogurt drinks), then the level of anxiety and aggression declined.

He therefore suggests being careful with antibiotics, because they intervene in the bacterial flora of the intestines.[29]

Food also influences the intestinal flora and is evidently able to have an effect on our mood. The intestinal flora can change rapidly depending on food: vegetables and fruit provide a favorable profile, meat without vegetables and fruit an unfavorable one. The current probiotic drinks seem to create a short-lived but not systematic favorable change.

The whole of the intestinal flora is called microbiome, and around it a booming science is growing. Microbiome appears to have great influence on all kinds of physical and psychic disorders. There is no end in sight to the discoveries in this field.

Consciousness, therefore, does not hide in our skull. We are not our brain, and the function of the body is not at all solely to carry, feed and propagate the brain. *The body makes a decisive contribution to consciousness. Consciousness exists in the entire organism.* We can trace this from the beginning of the evolution of the animal.

CHAPTER 9

The Brain and Human Freedom

The entire evolution of the living world is a movement from a lesser to a greater measure of freedom. This encompasses the essence of the evolution of the forms of life. The highest life form enjoys the greatest freedom.

Vasily Grossman[1]

A BBC popular-scientific documentary about human evolution focused specifically on the transition from Australopithecus to Homo, thus from ape-man to the human-like; this transition began some two million years ago with Homo habilis. I was surprised by a new insight; in an interview a scientist related that this transition began with the growing size and strength of the thumb and the possibility of opposing it to the other fingers. And now the important point: this was a step that was not the consequence of adaptation to the environment, but of new behaviour, namely more subtle use of stone tools, in the form of cutting. Before that time, stones had only been used for hammering. With the aid of pressure gauges it was shown that cutting with a sharp stone requires a great deal more strength in the thumb than hammering.

'A step that was not the consequence of adaptation to the environment, but of new behaviour.' Until recently, this sentence would have been completely rejected by neo-Darwinists (Darwinism expanded with the theory that evolution is the consequence of accidental mutations in DNA) if in the meantime such adaptation had not been demonstrated to exist without a doubt. Darwin's French precursor, Jean-Baptiste Lamarck (1744–1829) had already raised the possibility that inherited *behaviour* was the driving force behind evolution. However, this put him among the scientists who were later considered to be wrong. Today these are slowly being rehabilitated.

In most cases, evolution is not the consequence of accidental mutations that generate a better chance of survival, but it turns out that organisms *themselves* are able to adapt their genome, their DNA, either to changed *circumstances* or to changed *behaviour*. Accidental mutations usually have no effect, or they bring bad news: they produce genetic diseases. They rarely produce improvements. Anyhow, cells constantly receive signals instructing them to change something in their DNA. Genes are turned on and off. And this results in a change in the behaviour of the cell. The specialty that studies these phenomena is called epigenetics. But by itself this changes nothing in the composition of the genome (the total DNA sequence). When the signals become insistent and compelling by changes in the behaviour of the organism – whether or not caused by changed demands of the environment (which could be called a form of stress) – it may happen that turning on a certain gene has insufficient effect. This gene then can be *doubled* (duplication).[2] Genes may also change place, or even be recombined, on the chromosome. This cut, copy and paste function takes place especially in case of ecological stress, which makes it possible that the genome is rewritten so that it receives a different significance. It has been found that the place of a gene on the chromosome makes a big difference as regards activity.[3] In such a case, therefore, accidental mutations have nothing to do with it.

Genes constantly interact with the cell and with the entire organism. They can do nothing by themselves. The genome of every organism contains elements that can cause the above named changes to happen (duplication, transposition, recombination); these elements are called *transposable elements* or *transposons*.

The advantage the vertebrates built up in intelligence some 550 million years ago was ushered in by two duplications of the gene that codes for the creation of synapses. These were no accidental mutations. According to the authors of the articles I am here referring to, accidental mutations in these genes lead to psychiatric disorders.[4]

Human and Chimp

Another example of spontaneous genome changes as a result of specific stress in larger organisms is the following. Human beings have 98.7% of their genes in common with chimpanzees. As shown by the activities of transposable elements, the remaining 1.3% codes principally for the brain, senses and immune system. Generally speaking, humans have an advantage in these areas. For example, there is a gene (SRGAP2) of which chimpanzees and other mammals have only one copy but humans have four copies. It was therefore duplicated twice. One of its tasks is making synapse connections. Because of this, humans can have infinitely more connections in their brains.[5]

Only in regard to immunity for the HIV virus do chimpanzees have the advantage; they cannot have that illness. It has to do with the gene CCL3L1 that can prevent the connection of the HIV virus to a cell. While humans have only one to six copies of this gene, the chimpanzees have eight to ten of them. It has been found that newborn babies of an HIV-seropositive mother possess after birth a larger number of copies of the gene than their mother, and are consequently less susceptible to infection.[6] The transition from one (large) species to another (phyla) results invariably in a duplication of genes, sometimes even of an entire genome. Maybe you will find the following discovery of interest. Humans differ individually in their quantities of the gene CYP2D6, which plays an important role in detoxing toxic plant material and medications. Some people have three copies of this gene, so that many medications simply have no effect on them. Others may just have one copy and are therefore hyper-sensitive to most medications.

In 10% of western humanity the gene is completely lacking; the same medications can therefore lead to poisoning in these people.

Also in the further evolution of Homo sapiens there have been significant changes in the genome, not in response to ecological stress, but to behavioural changes. Humans do not adapt to the environment as animals do, but adapt the environment to themselves. This actually means that in the case of humans, we should speak of a very different kind of evolution. When agriculture developed and human beings began to digest more and more starch, the gene that codes for starch digestion enzymes was doubled multiple times.[7] And something comparable took place when people in different locations began to drink cow's milk. In them, independent of each other and in a short time, genes in different places in the genome were changed for the enzyme lactase. This means that in each of these groups of people a different gene was changed.[8]

Our brain cells contain many more transposons ('jumping genes') than other tissue. This results in the fact that our neurons all differ in their genomes. Each neuron has some 80 to 300 unique changes in a gene. 'It looks as if each cell is there for a purpose,' said neuroscientist Allyson Muotri[9] who discovered this. Such differences are always a consequence of challenges in the (cultural) environment. This means that we, certainly when still young, can always adapt to cultural changes, which we of course cause ourselves. And in turn, it means that our brains are most probably different from those of our ancestors. Actually, we might have suspected as much, for many parents are often surprised by how much more at ease their children are with modern technology than they themselves.

Natural creation

Just as we are not at the mercy of an unchangeable and dictatorial brain, organisms are not at the mercy of unchangeable and dictatorial genes. Recent developments in genetics and neuroscience show the same picture: *changing is something we do ourselves.* French philosopher Henri Bergson had already predicted it.

In his book *Creative Evolution* he wrote that nature itself is creative, driven by *élan vital*.[10] Organisms are purposeful; they try to survive by staying as autonomous as possible from the environment. The same survival instinct drives evolution toward organisms with ever greater autonomy. Thus autonomy not only appears to be the goal of an organism, but also of evolution.

Evolution with a goal? This idea has been a scientific taboo for a long time, because it reminds many people too much of creationism or intelligent design, both of which imply that the hand of the Supreme Being can always be recognised in the many forms and development patterns in nature. But what if it is living nature itself that demonstrates purpose? Then the story suddenly has a different character.

Encephalisation

The German biologist Bernd Rosslenbroich has put forward a theory about the major transitions in evolution[11]. What drove fish to become amphibians? And what drove them to become reptiles, birds and mammals? He describes organisms as beings that try to keep their order in spite of the entropy of the lifeless environment. To be able to do that they have to be autonomous. He sees that every major step in evolution is a step into more autonomy, more freedom. This he illustrates with all the anatomical changes evolution shows in these steps. We can follow his ideas in this way:

In evolution the first unicellular organisms that are supposed to be the ancestors of the animal kingdom called the protists, already react on their environment. We could call that protoconsciousness. Some of them even have eyespots, and can react on light, but they have, of course, no neurons, being unicellular.

The first multicellular organisms that possess nerve cells are the coelenterates: jellyfish, polyps, chorals, sea-anemones and other sea flowers, and comb jellyfish. The outer cell layer, the ectoderm, contains a network of nerve cells that function at the same time as sensory nerves. They are connected together in nodal points called ganglia. All impulses from outside are communicated to the entire body.

Stimuli can still go in two directions in these animals; later in evolution that changes. The body reacts automatically. Touch a polyp (hydra) and it immediately pulls back. It also does this when it touches anything. It still reacts rather like an automaton. It has no central nervous system; everything is on the periphery. There is, therefore, no brain with the potential of learning to reflect: 'Was I touched or did I touch something myself?' This does not mean that the polyp has no consciousness – its entire body is conscious. It perceives impulses and reacts to them, but it has but little freedom. The hydra may not be very intelligent, but that is amply compensated by a vitality much like that of plants. All its parts can regenerate when they are cut off.

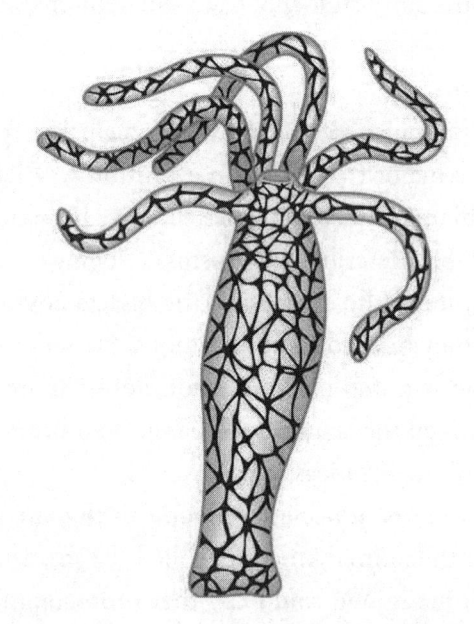

Figure 9.   The nerve system of the polyp with ganglia as nodal points

The Sea Slug

Neuroscientist Eric Kandel earned his Nobel Prize by his work on the learning capacity of the *Aplysiacalifornica*, a sea slug [12] This enormous slug of some 12 inches long has a well-ordered quantity of crudely built neurons (20,000 of them) and is a step beyond the polyp.

9. THE BRAIN AND HUMAN FREEDOM

When it is touched, it withdraws. But when this happens repeatedly in a non-aggressive manner, the slug becomes used to it and relaxes. In neuro-speak this is called *habituation*. Something entirely different can also take place. When we inflict an electric shock on the sea slug it shoots away, and when we do that repeatedly it does that even faster. It has become over-sensitive; this is called *sensitisation*. In that case it may even react over-sensitively to normal touch.

What is the difference with the polyp? Just as the polyp, the slug has sensory neurons, but these send the touch stimuli to a central point in the head which, however, cannot yet be called a little brain, but a set of somewhat larger ganglia (nodal points). From there, motor nerves lead to the periphery where motion is caused. The stimuli can therefore move only in one direction. Stimulation of the sensory nerves is translated in a ganglion into stimuli in the motor (movement) nerve cells, resulting in withdrawal or flight. Habituation weakens the connection between sensory and motor nerves in the ganglion, sensitisation strengthens them – sometimes even by making extra connections.

This is the manner in which the slug learns, says Kandel. In the slug also the whole body is conscious. Here too, the central ganglion is not the place where it actually feels the stimulus, but the place where it changes and 'learns.' The slug perceives the stimuli, but does not always respond in the same way. It will not have much inner awareness, and in this sense its reactions are automatic. And it will not remember anything consciously. Therefore it does not really look like our conscious learning, but it does reveal which mechanisms may be responsible for implicit learning or for automation. Kandelian 'learning' happens in the ganglion, which merely sees to it that there is a certain variation in the response to a stimulus. *Nevertheless, the central ganglia are thus a first neuronal step on the way to freedom, where between perception and (motor) reaction a space arises to learn to differentiate behaviour. Later, due to the expansion of the ganglia into brain, this learning capacity will be developed further so that there will be more space, and thus more freedom, between perception and reaction.*

This learning capacity exists due to neuroplasticity, which is the *core business* of the brain.

Sooner or later in evolution, brains appear in most animal species. The literature commonly emphasises the increase in brain volume, so-called encephalisation. Larger brains offer a larger chance of survival because they yield more freedom to respond to unexpected challenges.[13] The brain is an organ for which the old was not exchanged for something new in evolution. It is built up of old structures that have remained present, and from which new structures have proceeded and have been added. But of course it is interesting to examine what exactly was added. What functions were expanded? We can gain an impression of this when we look at the evolution of mammals in this way as shown in Figure C on the inside front cover.

In the rat the primary projection areas of the senses and the primary motor cortex take up most of the space in the brain cortex. These are, on the one hand, the areas where the stimuli of the senses come in first, unchanged, and on the other hand, the areas from where muscle movements are directed via outgoing stimuli. The basic form of the brain, the connection between centres of perception and centres of movement, can still be clearly recognised. In the cat there are some more convolutions, apart from the primary projection areas, and in the human being the space of the primary visual, auditory, motor and sensory cortex has been greatly surpassed by the neocortex, which has enormously expanded. But what is the function of this neocortex?

It turns out that the increase in brain cortex in both cat and human being can be completely attributed to the so-called association areas. Humans have many of those (see Figure E on the inside back cover). The association cortex has a synthesising function, in which pattern recognition plays a large role. New or previously learned patterns are added to the sense impressions that are processed in the primary cortex area, so that meaning can be given to them. Examples of this in the *visual cortex* are areas that have learned to recognise the direction of a movement, the degree of an angle, colour, a characteristic feature, faces, the relations and places of objects in space.

9. THE BRAIN AND HUMAN FREEDOM

This applies not only to the area of the senses. The primary motor cortex also has association areas (the so-called pre-motor or supplementary motor cortex) that do not control muscle movements but help plan the patterns of actions. The entire evolution of the brain can be viewed as an ever-progressing effort to be no longer dependent on what the senses bring in, and the corresponding inculcated responses. We witness an ever widening separation between perception and reaction. Something (physical) moves in between, something that makes possible refinement and differentiation in perception, a coherent image and, equally important, differentiation in actions. This takes a little time. In this time a free space comes into being for consciousness, a space to hold the reaction for a moment, with the possibility to deliberate what the best form of the reaction will be. In other words, it is a space to give meaning to sense impressions; and a space for control of actions, since an action can first be 'proposed' before it is executed. Space also arises for feelings that play a role in the question of whether or not to react, and to what extent. In all this it is important to realise that these meanings have to be *learned*; they are not inherently present in the association cortex.

In this way, *the brain forms the instrument of consciousness*. This makes the role of the brain in pain comprehensible. We are not polyps. We need the connection of pain perception with learning, imaging and deliberating to be able to know how to react, at any rate, as long as we are conscious. In Chapter 6, I described that when pain is inflicted during sleep the centres that are related to the pain localisation are activated in the same way as when this happens when we are awake. Only, as long as we are asleep we do not feel the pain. The only difference in brain function is that during sleep the middle of the *gyrus cinguli* (see cortex cinguli in Figure D), which is involved in orientation and avoidance behaviour, is not activated.[14]

Space is created not only for more differentiated and more complete perception of the environment (most animals merely perceive what is important for their survival), but also for more creative use of the limbs. This makes it possible to employ the limbs also for things other than locomotion, such as manipulating, holding and seizing food, the way predators, raccoons, squirrels and hamsters do, and for handling primitive tools by non-human primates.

This means that the brain pre-eminently forms the organ that makes an ever-increasing freedom possible in evolution. Absolute freedom in the use of tools and instruments is created by the hands of the human being. Because of our erect posture, these are no longer needed for locomotion. As a result of this and their special construction, ever since Homo habilis the hands can be employed as instruments in any conceivable manner. This mutual influence of head and hands has resulted in an enormous degree of freedom for the human being.

About the Brain

The human brain differs from those of 'non-human mammals' in roughly two aspects. In the first place, the brain cortex is thicker because it consists of more cell layers, but it is also larger and more important due to the association areas. This enables us, much more than the animals, to give *meaning* to sense impressions, and act creatively. What distinguishes us from the animals more than anything else is that we are able to think something is beautiful or ugly, true or false, and judge a deed morally. Plato called it the beautiful, the true and the good.

The second aspect is the importance of the frontal lobe. This is above all due to the increase in direct connections from the frontal cortex to the rest of the brain. The association areas in the frontal lobe, the pre-motor cortex, give us incredible freedom of movement as compared with that of animals. No matter how capable animals may be in their specialisations, they can't come close to the diversity humans demonstrate in the Olympic Games, in the concert hall, or in a trade. Moreover, the frontal cortex helps humans not to react automatically, so that they can take time to deliberate what action is appropriate in a situation. This area makes it possible that reactions to impulses from the sense organs and impulses from the lower brain areas do not run automatically, but can be either stopped or permitted.

9. THE BRAIN AND HUMAN FREEDOM

Here again, free will is made possible by control of action. We can exercise this self-control thanks to our possession of the extensive frontal lobe. In a dog this lobe takes up 7% of the entire brain volume, in monkeys it increases to 17%, but in apes and humans it is 35%. (The visual cortex takes up an additional 30% so that only 35% is left for the remaining cortex.) What do apes do with this? Much less than we do, we know that. This is of course connected with the fact that apes have smaller brains than humans, so that their 35% is less than the 35% in humans. In addition, humans have much greater gene expression, which makes a far more extended range of neuronal reactions possible.[15] But what is undoubtedly at least equally important is the fact that apes have much fewer nerve fibres, fewer connections, but especially less myelin. Myelin is the white substance around the nerve fibres, one of the effects of which is that impulses travel through the nerve much faster. In humans, myelin formation begins after birth and is only complete around the twenty-fifth year of life.

In primates mirror neurons have evolved that make imitation easier. As was discussed in Chapter 3, this has enabled humans to develop language and culture. Culture offers much greater freedom than biological evolution, namely freedom from the rest of living nature. This a step ahead of what life did in the first cell, namely bringing freedom vis-à-vis *lifeless* nature.

Be that as it may, the fact that the brain carries with it the evolutionary heritage results, on the other hand, in a limitation of free will. For our nature is in large part defined by evolution, heredity, culture, gender and environment. However, the interesting thing in human beings is that, when they have grown into adulthood, they can learn to free themselves to some extent from these limitations.

Culture and Idea

All this evolutionary increase in freedom looks purely like the result of survival instinct. Is this space for freedom then a sufficient explanation of human culture? Does freedom lead to art and science?

Or to morality, love, wisdom or meaning? Does the brain itself, somewhere between the input from the senses and the output of action, create motives to rise above nature? Since the first Homo sapiens some 200,000 years ago it took another roughly 160,000 years before the first forms of art – sculpture and cave drawings – emerged. It did not happen all by itself.

In the case of morality, however, things were different. Morality is not an invention of the human being, as was demonstrated by Frans de Waal in his book *Good natured: The Origins of Right and Wrong in Humans and other Animals*. There are plenty of examples of altruism among animals, although the motivation for it remains unclear. Do animals know empathy? Certainly. And the fact that they know altruism is fortunate, otherwise to be a decent human being we would have to wage a constant battle against our 'animal' nature. No one could keep that up for very long. Some people even maintain that animals have feelings of justice, but this is limited to the indignation of a test animal when it feels shortchanged compared to another test animal. Others even ascribe culture to animals; this turns out to be the development and copying of new habits.

But culture is more than habits. There is an idea behind it. Humans form ideas. The fact that we had, and have, an idea that life can be better is probably the explanation of how humans have become cultural beings. The same goes for morality. Being a moral entity does not mean acting morally all the time, but having an *idea* of what correct moral acting is. How did people develop ideas on this? We don't expect such ideas among animals. Are they produced by the brain? Except for greater volume, there are no new structures by which human brains distinguish themselves from those of other primates. Why then would human brains suddenly want to rise to a higher level? Could it be perhaps that there is an essential difference between humans and animals that is not caused by biology but by consciousness?

9. THE BRAIN AND HUMAN FREEDOM

Consciousness Revisited

The above may be a general description of the central nervous system, but we still have the riddle of how conscious experiences come about. Consciousness exists already in a primitive way in single-cell organisms. But do these organisms experience anything? *What brings about experiences?* Evolution does not tolerate anything that has no purpose. Brains use up a lot of energy, a large part of which is used for the operation of nerve cells that are deemed to be involved in experiences. Experienced consciousness must therefore have a purpose in a biological sense.

However, neurodeterminists tell us that experienced consciousness plays no role whatsoever in our behaviour, because something that is not material or a force of nature is supposed to be incapable of influencing matter such as brain tissue. This would mean that such consciousness has no biological purpose. Many organisms are doing just fine without it and can spread their genes perfectly efficiently by unconscious, automatic behaviour. In this regard, human (self) consciousness is actually a handicap, for we can make mistakes and have doubts. By the intervention of consciousness, sexual partners can choose or reject each other on grounds other than genetic fitness. Thus we can change our question: *Why does experienced consciousness come into being?*

The theory of the selfish gene is no help here. Not until we can decipher the roles of autonomy (and therefore freedom) and survival instinct can we grasp the enormous step evolution made in the puzzling development of experienced consciousness. Biologically this step is indeed extremely useful. Every little bit more consciousness results in an increase in autonomy. This makes the importance of the evolutionary growth of consciousness comprehensible, and perhaps also the even more puzzling, yet inevitable, development of human self-consciousness. For this opens the door to the observation of our own consciousness, so that we may direct it.

Maybe this is the answer to the question: why is one animal's consciousness deemed superior to another's? The point is the degree of freedom, the feeling that we have to do with a more or less individually acting entity. There is a word for this: agency, the power to choose to act. With a mosquito we don't really get that feeling, but it begins to rise with the vertebrates, certainly in the case of the more evolved ones. In humans it even becomes *essential* that we have the feeling of dealing with an individually acting entity. If that is not the case, something is wrong. If free will exists, the human being is certainly the entity with the greatest chance of possessing it. But free will is not something 'out there;' it is always the free will of someone, of a 'free willer,' a person, a subject, a self, an 'I.' For after all, if it were the free will of a brain, it would not be *free*, but *determined*. When we follow the picture of evolution described here, we will understand that the existence of the free will, and of that individual entity, the 'free willer,' the self, is actually pretty obvious. Why do so many neuroscientists nevertheless deny the existence of free will?

CHAPTER 10

No Free Will in the Lab

Do I believe in free will? I have no other choice!
Isaac Bashevis Singer

To the question of the meaning of an action physiological or neurological causes are completely irrelevant.
Ludwig Wittgenstein[1]

Science journalist Mark Mieras signed up to be a test person in a neuroscientific experiment because it fitted perfectly with his objectives of the last few years. He wanted to show what insights the neurosciences bring us, and therefore gladly offered himself as a test subject. Now he has already written three books about his experience and devoted many articles, radio talks and lectures to it. This particular experiment was to provide insight into the existence of free will.

It was to be a variation on the famous experiment of Benjamin Libet of 1983,[2] in which test subjects had to bend their wrists or lift a finger and, while watching a clock showing milliseconds, fix the moment when they decided to perform that action. Mieras wanted to write an article about the new experiment.[3] But before he went to the brain lab he first had to take his children to school in a carrier-tricycle (a big tricycle with a cargo area in front). He wore a helmet because he knew how vulnerable the skull is and wanted to avoid non-congenital brain damage. This was how I met him in the Vondelpark in Amsterdam. We exchanged a greeting.

After delivering his children at school, he went to the laboratory of the University of Amsterdam, where he put a cap with electrodes on his head with which his brain activity was to be measured. The experiment in which he would participate could be called a test of 'free won't' he had to stop an activity spontaneously. The activity was the following: with the aid of a computer mouse he had to imitate the horizontal movement of a cursor on a computer screen with a little ball. A few days before he had had a practice session in which he had to learn to stop 'spontaneously,' because it appears that this is even more difficult than spontaneously beginning something. For instance, as he wrote in the article, he was not allowed to count to three and then stop, 'as I did as a child at the jump-rope when I tried to decide when I would jump in.' It had to be really spontaneous: 'I had to surprise myself. When I really gave it all my attention it seemed almost impossible. No matter how spontaneously I tried to decide, there were always thoughts or feelings in my head preceding the decision, as if every decision is imposed on us from higher up.'

The resulting data demonstrate that seconds before every 'spontaneous stop' measurable activity takes place in the pre-motor cortex. 'Evidently this part of my cortex knew well ahead of time that a 'spontaneous decision' was coming and was already preparing for it,' wrote Mieras. This is exactly what Libet also found in his data, and he explained it in a comparable manner. He said that before the decision a 'readiness potential' can already be measured that proves that the brain unconsciously prepares for the conscious decision. He also measured the timing: in his experiment the readiness potential arose an average of 550 milliseconds, and the conscious decision only 200 milliseconds before the hand movement.

There are even stronger examples of preparation in the brain. A YouTube video with the title *Neuroscience and Free Will* (part of a BBC documentary) shows a presenter, Marcus de Sautoy, who subjects himself to an experiment in an MRI scanner performed by neuroscientist John-Dylan Haynes. He is given a little box with a button in each hand. He may decide which button to push. Afterward, Haynes shows him which area already tells *six whole seconds* ahead of time which button he will push. At the end of the video the presenter collapses dazed on the steps of his laboratory and wonders whether he is really an automaton. By the way, what is not reported is that Haynes was able to predict the right outcome in only 60% of the test cases.[4]

In 2008 Chun Siong Soon and the same J.D. Haynes published a letter in the journal *Nature Neuroscience* that created a stir because it said that with the aid of fMRI scans it had been discovered that, before the readiness potential in the pre-motor cortex, there is already measurable activity in, successively, the prefrontal and parietal cortices.[5] This happens a full ten seconds before the movement is made. Evidently, the authors conclude, the decision is made in the pre-motor cortex not 350 milliseconds before becoming conscious, but much earlier, way up front in the brain. If we now continue to believe in conscious free will, we have to ignore all these facts.

Walk Your Talk

If that is true, we must accept it. However, acceptance creates a problem. Remember the window kicker from Chapter 1? At that time I did not yet know that the life motto 'There is nothing I can do about it. I did not make myself, did I?' and the behaviour of the window kicker might have a great deal to do with each other. According to a study done in Minnesota and California in 2008, people who do not believe in free will more often display antisocial and irresponsible behaviour.[6] Some of the test subjects were given Francis Crick's text we have cited before:

> You, your joys and your sorrows, your memories and ambitions, your sense of personal identity and free will, are in fact no more than the behaviour of a vast assembly of nerve cells and their associated molecules.[7]

Then they had to solve mathematical problems for which they earned money for each correct answer. They were asked not to cheat, although that was very easy to do with a little computer trick. Among the ones who had read Crick's text 60% more cheated than among the others. In a follow-up study in which they could also earn money the readers of Crick's text stole significantly more than the others.[8]

Although this result confirms my prejudices, I still have my doubts about the study. The reason is that virtually all psychological and behavioural studies (96%) are done on mostly male, beginning psychology students in western, especially Anglo-Saxon countries in, of course, western universities (99%). Canadian anthropologists and psychologists have called these test subjects western educated industrialised rich and democratic (WEIRD).[9] This group has been shown to differ substantially from the rest of world population in behaviour, views and psychology. What they demonstrate is commonly extrapolated to all of humanity as if they revealed inborn natural laws in universal human behaviour. But since the laws thus discovered are often not found in other population segments – remember the lines of equal length of the Müller-Leyer illusion – it turns out that in this case we are not witnessing general human behaviour, but culturally determined, therefore learned behaviour.

Professor Dan Ariely made the study of everyday forms of dishonesty his life work. He also studied the reverse, namely whether one can foster honesty. In an experiment in which part of the subjects had to write down as many of the Ten Commandments as they could remember, this group accomplished the test in complete honesty, in spite of the fact that cheating had been made easy.[10] There were many more ways to foster honesty, such as a mirror in which the participant was confronted with himself, a signature under a contract stipulating honest behaviour, and so on.[11] Evidently a slight confrontation with one's own (ideal) image is indeed able to make a difference in behaviour.

In another study, temporary employees were asked whether they believed in free will.[12] The scores were compared with their work performance. Those who did not believe in free will were generally less social in their behaviour, less consistent, and achieved less in their work. According to another study, lack of belief in free will leads to increased aggression and less readiness to help.[13]

Keep Them Dumb or Tell the Truth?

Disbelief in free will therefore does not seem to be good for one's character. At the end of the article describing that reading Crick's text – which contains the statement that the brain determines everything – can cause antisocial behaviour, the authors propose that if exposure to deterministic messages causes unethical behaviour, we need to look for ways to safeguard the public against this. Does this mean that we should not publish this unwelcome conclusion? In his time, eighteenth century French philosopher Voltaire made a similar suggestion. He thought that the common people should be kept ignorant of the view developing among the 'enlightened' elite that, after the creation, God had lost interest in the world. The people would then undoubtedly start to misbehave.

To adapt scientific research to what is socially desirable does not seem to me a good idea. You can only embrace that if you really think that the public lacks the freedom to mistrust the message. But to check whether the deterministic message makes sense seems to me totally sensible. For perhaps Crick, Swaab and Lamme are simply wrong in their conclusions. Actually, the experiments themselves described here provide us already with an indication in this direction, for when our view of life influences our behaviour, doesn't that in fact support the idea that we can control our behaviour to some extent? If the researchers had recognised this and reported it, they would not have had to make his strange suggestion.

But there are additional reasons to be suspicious of the idea that free will does not exist. For what do these experiments really show? Does pressing a button actually have anything to do with free will? Is free will really about such silly movements? Doesn't the question of free will rather relate to 'meaningful' actions and decisions?

Readiness Potential Is Not the Beginning of a Movement

And what does this 'readiness potential' mean that is evidently observed 'in the head'? Psychologist Christoph Hermann was intrigued by this. In an article that appeared a few months before that of Soon and Haynes, he showed that readiness potential has nothing to do with the action itself.[14] He gave test subjects two buttons to push. Which button to push was made clear by a sign that appeared on a computer screen. Indeed, the action of pressing the button was preceded by 'readiness potential.' Only, this occurred already before the sign appeared on the screen, meaning: before the test subjects knew which button they had to choose. But the readiness potential did show the general expectation that a decision had to be made soon, as if the brain were bracing for it.

Libet also described that his test subjects — just like Mark Mieras who felt it 'in his head' — reported experiences that indicated a kind of general preparation or planning to do something in the future. Looking back and rereading Libet's original article, we can read that 'readiness potential' also occurred with test subjects who did not lift their finger! Maybe the pre-motor cortex doesn't yet 'know' anything, but is switched on to be ready for whatever is needed. The activity that Soon and Haynes witnessed ten seconds in advance in the frontal cortex therefore probably has nothing to do with the eventual movement. For this reason the term 'readiness potential' seems to be well-chosen. Mark Mieras even calls it the 'vestibule of free choice,' with which he makes clear that, in his view, this unconscious brain activity proves nothing against free will. Justifiably so, for *in whom does this brain activity take place?* In his article Mieras therefore concludes that free will is a philosophical question, not a neurological one.

Unlearning Free Will

As mentioned by Mark Mieras in his article, his experiment was repeated by David de Rigoni. Here also part of the test subjects had to read the same text by Crick that appeared to lead to antisocial behaviour: 'You, your joys and your sorrows, your memories and ambitions, your sense of personal identity and free will, are in fact no more than the behaviour of a vast assembly of nerve cells and their associated molecules.'[15] The experiment showed that reading this quotation had a conspicuous effect on brain activity. The ones who had in this manner been talked out of belief in free will showed much less 'readiness potential' or, in Mieras' words, less 'vestibule of free choice: is someone who does not believe in free will for that reason left with less free will?'

Disbelief in free will thus not only results potentially in antisocial behaviour, but also in different behaviour by neurons in an action. Rigoni concluded: 'The idea that they are not responsible for their behaviour seems to lead people to a diminished experience of this behaviour as being theirs'. Follow-up research seems to indicate that the conviction of having no free will results in the brain in an increase of automatic behaviour. Daniel Dennett seemed to surmise this already in 2003: 'Perhaps there are two kinds of normal people: the ones who do not believe in free will and *therefore* have no free will, and those who do believe in free will and *therefore* indeed possess free will.[16]

This is exactly what the angry patient of Chapter 1 showed, the man who said 'I did not make myself, did I?' It seems to me, not only because of the broken window, that this is not really a desirable situation.

Causal Chains

Maybe there was something conspicuous about the first paragraph of this chapter. It is an account full of decisions. Other than the decision in the laboratory, which was the centrepiece of the whole story and showed the usual causal sequence reigning in physics, all the decisions taken in 'real life,' before Mieras arrived in the laboratory, went against the direction of the causal chain, namely not to the past but to the future. It may be true that intentions in normal life can at least in part be accounted for from the past, but they also invariably point to the future. In reality, we all make pictures for ourselves all day long of *the* and, most of all, *our* future, and we try to manage things in our interest. That is what free will ought to be about.

But of course this does not hold true for the agreed 'decision' in the laboratory, for it does not refer to anything and has no context. Mieras was not even allowed to count, something he would evidently have done in normal life. The source of our behaviour does not at all lie exclusively in the stimuli that immediately precede it. 'When we want to make voluntary actions seem involuntary, the first thing to do is to strip away their context … and then effectively break them down into their physical elements,' said geriatrician Raymond Tallis.[17] This is only possible in a laboratory.

It is striking to see how few problems neurodeterministic thinking has with reducing the concept of 'free will' to pressing a dummy button that doesn't accomplish anything. Actually, this experiment is not about free will at all, but about arbitrariness. Every decision Mieras made before the experiment was motivated. Oho, our determinists may say, motivation points to a cause in the past and is therefore not free. On the contrary, motivation points to the future, a different future from the one that may be expected out of determinism. But the future does not yet exist and can therefore, viewed deterministically, have no influence. For this reason, the future also plays no role in Libet's experiment.

Self-Control without Brain Activity

In his test, Libet also included the possibility to revise the decision to make the hand movement before it was executed. This 'veto,' as he called it, occurred 100 milliseconds after the conscious decision and, therefore, also 100 milliseconds before the movement would have been made. Libet concludes that evidently there is indeed a free will in the form of the veto, *for that veto was not preceded by brain activity*. That is of course extremely interesting. For it means that although an action, for which a physical reason can always be found, presupposes prior brain activity, this does not hold true for control of the action. Libet therefore concludes that free will is not so much about impulsively doing what we feel like, rather it is about our control over our deeds.[18] Elsewhere Libet also says that, similarly to what was pointed out in the previous chapter, consciousness is not a physical phenomenon and can yet exert influence on the brain.[19]

It is noteworthy that both the 'automatic behaviour' and the 'veto' were hardly or never preceded by brain activity. Does this mean that the brain plays no role in 'arbitrary' actions? At least equally interesting is that this also demonstrates that our consciously acquired views make a difference in our actions. At first sight, freedom of action is then limited by view of life. But in normal circumstances, this view of life has been acquired consciously and freely. Is this, and I reiterate, not also a form of free will? And a source of motivations? In this way, most of what we do – except what we do by instinct such as sleeping and most of our eating – was at one time consciously imprinted.

Daniel Dennett, a materialist and thus a determinist, points out that the fact that we can learn has the consequence that we can choose our conduct, and can imagine the future for ourselves. This enables us to influence the future by our behaviour which is, in his view, the essence of free will In 2012 Dennet was interviewed in Holland about the denial of free will by the Dutch scientists Swaab and Lamme. 'Bizarre! Absurdly poorly substantiated! Logical errors one would expect of a student!' was his reaction. 'Your Swaab and Lamme figure in my rogue gallery of scientists who got things wrong.' [20]

Dennett's friend, mathematician and consciousness researcher Douglas Hofstadter, considered that although perhaps we do not possess free will, in any case we do have 'free won't.'

'Free Won't'

The notorious study of free will and obedience made by Stanley Milgram in the 1960's was a test of this 'free won't.'[21] Test subjects had to give ever stronger electric jolts to 'students' who had to repeat certain words in a 'memory experiment.' The students were actors and the electric jolts were not real, but the test subjects did not know that. Starting with the suggested 135 volt the students initially reacted with groans, then with cries, after 300 volt with banging on the wall and eventually with silence. When a test subject wanted to stop he or she was told that it was absolutely necessary for the experiment to continue. Many continued to 450 volt. But some of them refused to do so well before that level. This could tell us much more about free will than pressing a dummy button.

In 1971 a prison experiment in Stanford showed how difficult it is to resist authority, and also peer pressure.[22] Students were split into two groups, a group of 'guards' and a group of 'convicts.' After some time the guards started to abuse their power and the convicts began to behave submissively. The aggression of the guards grew worse and worse. Convicts were sprayed with fire extinguishers, had to undress in public and were treated like animals. Behaviour in a group can degenerate by imitating and outdoing each other. Social psychology calls this de-individuation, a form of loss of individuality, loss of 'I,' under the influence of peer pressure. It results in doing things one would never do on one's own. Also, under such peer pressure, one evidently experiences one's behaviour less as one's own.

There was only one student, Christina Maslach, whose task it was to interview the test subjects, who demanded that the experiment be stopped, while at least fifty outsiders were allowed to observe the events, none of whom had uttered any protest. Christina Maslach was one of the few who possessed 'free won't.' It would be interesting to study under what conditions this important phenomenon can occur, because 'free won't' here appears to be intimately connected to morally proper behaviour.

Self-Control

It seems to be an uncontested fact that not everyone receives the same amount of free will at birth. The Stanford marshmallow test was an experiment made with six hundred children in 1972. Young children were put in a room each with a marshmallow in front of them. They were allowed to eat it, but if they managed to wait fifteen minutes until the experimenter returned, they could have two.[23] Twenty years later the now adult participants were tested again. It turned out that those who had been able to control their impulses as a child were significantly more successful in later life.[24] In a research project among students[25] it was shown that higher self-control correlated with higher grades, less alcohol abuse, fewer food binges, higher self-worth and acceptance, and more success in relationships.[26]

When the marshmallow test was recently repeated, it was shown that the result of the temptation depended greatly on previous experiences of the children. When in the past promises had been made to them that were not kept they did not hesitate long to eat the marshmallow. The second treat probably would not come anyway.

It can also be done differently. Terri Moffit and Avsalom Caspi, who have done much research into the influence of the environment, events in life, and genetic influences on the occurrence of anti-social behaviour and psychopathology, discovered that self-control was a factor which, independent of 'nature or nurture' seemed to have an influence on the course of life. They approached it differently from the then denounced marshmallow test.[27] They made use of a longitudinal study in the little town of Dunedin, New Zealand, where 1,037 children born in 1972 and 1973 were tested every other year starting at age three. In 2015 they were aged forty, and 96% of them still participated. On each occasion they underwent a series of tests while their parents, brothers, sisters, school staff, colleagues and friends were asked questions about the participant's degree of impulse- and self-control.

The result was in line with the marshmallow test, but better substantiated. It was found that a low degree of self-control determines virtually every aspect of life in a negative way: addictions, criminality, high blood pressure, overweight, and financial problems. A high degree of self-control leads to higher levels of education and income, fewer divorces, and better health. Pictures of the test subjects at age 38 even show that those with less self-control age faster than those with more self-control. This is independent of social class, the self-control of others in the family, or intelligence.

In a test of five hundred British twins, Moffit and Caspi found that self-control is also not genetically determined. A hopeful aspect is the fact that some of these people acquired better self-control in the course of time and that they scored better in the factors mentioned. Moffit and Caspi therefore suggest letting children practise self-control. They propose to let them save money in different jars for various purposes, including presents for others.[28]

Thus, free will has to do with self-control and not at all with arbitrariness. Arbitrariness is not the free will we should 'will,' agrees Daniel Dennett, who has no problem combining a determined world with free will.[29] Arbitrariness, he argues, leads merely to misery.

How Smart is the Unconscious?

All neuroscientific objections to free will come down to the view that our actions are unconsciously prepared by the brain. The fact that this can turn out to the good has been shown by psychology professor A. Dijksterhuis.[30] He demonstrates that our (unconscious) feelings are often more useful for making decisions than our reason. I think this is right. He further contends that our actions, our behaviour, are unconsciously motivated, and that we formulate our consciousness of these only afterward in thoughts or expressed in words.

For neurodeterminists this has to mean: our brain is smarter than we are. A sentence to ponder deeply. For is that actually possible? At any rate, it isn't if we are our brain. And it is remarkable that most of Dijksterhuis' further examples illustrate the stupidity of the unconscious. For it has been shown that people's behaviour can be influenced without their being aware of it (priming). The unconscious can be fooled. That is nothing new, you will say, for why would we otherwise have advertising and propaganda? But there is much more to it. We can make people walk more slowly by letting them read a text about aged people, or give smarter answers to quiz questions by letting them identify with professors. We can make them fall in love by putting them together in an extremely demanding situation, and we can let them choose the best object from three identical ones in a row, of which the one on the right virtually always turns out to be the favorite based on convincing arguments.

This seems to indicate that we spend our lives just following the unconscious, automatic decisions of our brains. The first objection against this conclusion from the behaviour research on which Dijksterhuis, and also Victor Lamme (to be discussed later), base themselves is that this research also is of short term actions and decisions. It is never about career choices or opinions on important life questions in the framework of the kind of life people try to create for themselves. But what does become clear is that in daily life we are often happy to let our unconscious make our decisions for us.

For that matter though, it is interesting that Dijksterhuis does not think that we have no free will. After all, he said in an interview, the unconscious is also ours, and it is of a completely different quality than the unconscious of animals.[31] But also for him consciousness is unimportant: 'It just babbles away.'

Dijksterhuis shows many convincing research examples in which a large percentage of the test subjects almost automatically let their (short term) behaviour be influenced without having any notion that this is taking place. But what interests me is what distinguishes the small remainder of the test subjects who did not react in the same way from the others. Evidently, these people do possess free will in the sense in which Libet viewed it, namely control over one's deeds. Do these people perhaps have something to offer for which we would all want to strive, such as a considerable measure of self-control?

Ego Depletion

In any case, individual self-control seems to be related to the following. Free will in the form of control of our deeds takes energy, and when we have to control our actions too often and too intensively our store of energy becomes exhausted. Psychologist Roy Baumeister and his colleagues speak of ego depletion.[32] Thus they surmise, justifiably so in my opinion, that control of our deeds is a function of our *self* (*self-control*). They seated people at a table with a dish of radishes and a dish of freshly baked chocolate chip cookies. Half of the participants were allowed to eat only radishes, the other half only cookies.

During the time they had to wait for the (supposed) subsequent phase of the test, they received puzzles to solve, ostensibly to compare their capacities with that of students. In reality, the researchers wanted to study how long the participants would try to solve the puzzles to which, by the way, there was no solution. It turned out that those who had not been allowed to eat the cookies spent much less time trying to solve the puzzles than the others. Apparently, according to Baumeister, the self-control required to stay away from the cookies had undermined the total capacity for persistence.

Another cause of ego depletion is a low level of glucose in the blood. Fortunately, rest will restore this. A good night's sleep leads to renewed capacity for self-control, and most people are better at it in the morning than later in the day.[33]

Therefore the self-control to stay away from the cookies is perhaps not the only reason for loss of energy. The cookies themselves quickly bring glucose into the blood, which is transformed into energy in the brain. At any rate, the conclusion holds true that the persistence required for solving the puzzles took energy. The same phenomenon has been observed in cases of difficult choices and complex mental tasks such as logical reasoning and working with abstractions. Things like this take much more energy than easy tasks and automatisms. 'Put another way, ego depletion makes people stupid in complex ways but leaves them intelligent in simple ways.'[34] Baumeister has traced the practical consequences of these discoveries in his book *Willpower*.[35] The best way not to lose one's willpower for important decisions is not to waste time and energy on little daily choices. We never stop to consider whether or not we will brush our teeth in the morning. And that is a good thing too, because all these kinds of unimportant decisions together could result in ego depletion. Force of habit does not use up decision forces, so that we can use these for more important things.

Recently, a research project was reported in which no ego depletion was shown. Test subjects preserved enough motivation to complete a last task correctly.[36] What remains true, however, is that difficult cognitive tasks take energy, as Kahneman shows in the following section.

10. NO FREE WILL IN THE LAB

System 1 and System 2

The difference between automation – things we do unconsciously – and (self)control – things we do consciously – has been studied by psychologist Daniel Kahneman. He received the Nobel Prize for it. Before discussing this we have to touch on the concepts conscious and unconscious again. The unconscious is also a form of consciousness, certainly if we attribute consciousness also to animals, as I argued in the foreword. But it is a little strange to speak of 'unconscious consciousness' and 'conscious consciousness. The problem did not yet exist when we called consciousness *psyche* or *soul*. Since Freud we have known that the soul is unconscious to an important extent. It also becomes clear in the triad conceived by neuroscientist Joseph LeDoux as the content of consciousness: cognition, emotion, and motivation. Or, in other words, thinking, feeling and the will. Thinking is conscious, feeling half conscious (sometimes we have to make emotion conscious) and will, motivation as shown by social psychology, mostly unconscious. The solution, as we already suggested: it is all in the mind.

Kahneman has devoted his whole scientific career to the relationship of both forms of consciousness to each other, and prefers to speak of system 1 (the unconscious) and system 2 (the conscious). System 1 provides us with automatic solutions to problems which in a far distant evolutionary past were sensible. It gives us the 'first impression,' associative thoughts and ideas about causality that may or may not be right. But we can consciously update the intuitive capacity of system 1 by gaining expertise in the areas for which we have to find solutions. Only then do we possess a 'smart unconscious,' at least in the area of our expertise and skills.

Kahneman described his life work in his book *Thinking Fast and Slow*. Thinking with system 1 is fast, with system 2 slow. System 1 is automatic, gullible and takes no effort; system 2 is well-considered and takes energy. The self-control which, according to Baumeister, leads to ego depletion because of energy loss and has to be kept up with glucose, therefore belongs to system 2.

To save energy we prefer to use system 1 when we make judgments or come up with solutions or do something without thinking. We are lazy.

A shocking example is his study among eight Israeli judges who had to decide on the early release of Jewish and Palestinian prisoners.[37] There were many applications and the decisions had to be made quickly. The standard decision was rejection. Still, it appeared that the decisions on prisoners with comparable records were not always alike. What was the cause of the differences? No, not the Jewish or Palestinian origin of the prisoners. It turned out that the quantity of approved applications peaked after the meals and snacks and then diminished to zero shortly before the next snack.

As Kahneman also discovered, our pupils increase in size and our heartbeat accelerates when we have to concentrate our attention or make a mental effort (system 2). 'Just like the electricity meter, the pupils of our eyes give us an indication of the speed at which mental energy is being consumed.' When we try to solve difficult sums by head, our pupils grow larger. As soon as the person inwardly gives up, the pupils become smaller again, so that Kahneman could then ask a test subject why he had given up. The subject would then be astonished that Kahneman knew that.[38]

By the way, our pupils also grow larger when we are suddenly interested in someone or something. For that reason it is best to put on sunglasses when playing poker or when we want to pretend lack of interest in a negotiation. System 2 cannot handle too much simultaneous information – see the 'invisible gorilla' in Chapter 6. System 1 does that better, unconsciously of course. Visual illusions such as the Müller-Leyer illusion can be ascribed to system 1.

Kahneman gives a few examples of automatic activities of system 1:

- Seeing that an object is farther away than another object (source of possible visual illusions);
- Detecting the source of a certain sound;
- Completing the expression 'war and ...';

- Showing revulsion when seeing a frightening picture;
- Detecting hostility in someone's voice;
- Solving the sum 2+2=?;
- Reading texts on billboards;
- Driving a car on an empty road;
- Understanding simple sentences.

Most things therefore take their course automatically in our consciousness. Frequently this is fine, sometimes even better than what conscious rationality can offer us. But it can also easily go wrong. System 1 can be fooled by the silliest stimuli, as Kahneman gives away: Physicians were given the choice of two treatments of lung cancer: surgery or radiation. One group was told that surgery had a 90% survival rate, while the other group was told that it had a 10% chance of death. In the second group many more physicians chose radiation. Physicians get a fit at the words death and dying, and it makes them irrational.

No less shocking than the test of Israeli judges is the example Kahneman gives of a test with German judges with more than fifteen years of experience.[39] They had to determine the right penalty on a (imaginary) woman who had been caught shoplifting. They had all received one of two doctored dice that been so manipulated that they always came up with 3 or 9. The question they were asked was whether the woman should be sentenced to more or less than the number that came up. The judges who threw 3 chose for 5 months, and those who threw 9 for 8 months. Experts too can be confused by irrelevant information.

Choices are rarely made on purely rational grounds. For it takes energy. In such cases it is evidently not at all smart to rely on one's unconscious. It is really not a bad idea to spend some conscious time on an important decision, and to eat something. However, the fact that we rarely do this still does not mean that we are merely predetermined automatons, as Victor Lamme contends in his book *Free Will Does Not Exist.*[40]

The Course of World War II Was Set

Cognitive neuroscientist Victor Lamme is a declared neurodeterminist. By his version of a decision by Winston Churchill that was to determine history, he attempts to meet the objection that research into free will is merely about influencing short term behaviour. The decision involved the sinking of a French warship that threatened to fall into German hands in an Algerian port. It resulted in 1297 casualties and 350 wounded, all French. Churchill himself related that he had weighed all pros and cons, and that this decision certainly demonstrated to the world that he was serious about the war against the Nazis.

According to Lamme, the decision had everything to do with the fact that 25 years earlier Churchill and the English navy had suffered defeat against the Turks, because their French allies left the scene at the critical moment. Lamme knows better what Churchill's motives were than Churchill himself, not a weighing of pros and cons but a combination of inherited characteristics, training, rewards, disappointments and other experiences, all resulting in antipathy towards the French. In brief, the decision was predetermined, not a conscious choice, not a question of free will. And Churchill had not the least notion what his decision was based on.

Did Churchill really not have the opportunity to choose differently? Suppose he had, history might then have taken a different course, who knows? But what in any case would remain the same is Lamme's commentary. In Lamme's story only the decision itself would have to be changed, but the rest – a combination of inherited characteristics, training, rewards, disappointments and other experiences – which describes the motives, could for good or bad reasons remain the same.

These kinds of statements once caused science philosopher Karl Popper to demand that statements which pretend to be scientific must have the characteristic that they can be contradicted (called: falsifiability). A statement is only scientific if it can be contradicted. Unfortunately, Lamme's story about Churchill fails in this regard. And the same holds true for the notion that a researcher knows better what someone thinks than the person himself. It means – and I can't find a better way of saying it – that we are witnessing pseudo-science.

Lamme calls Churchill's story about his considerations the product of 'the chatterbox,' Churchill's consciousness that talked rubbish by saying that he had weighed the pros and cons, but did not realise how matters really stood. Of course, Lamme himself is never bothered by any such chatterbox.

Prefer an Automaton?

The most important message of Lamme's book is that we are automatons, and the reason this is not immediately obvious is because of the role of the frontal lobe. There is a neurological syndrome in which people indeed seem to act like automatons. Lamme describes the work of neurologist François Lhermitte who examined people with injuries to the medial frontal brain cortex.[41] These patients have a compulsion to pick up any object in their field of vision and use it for the purpose for which it was made. Lhermite calls this 'utilisation behaviour.' A pair of spectacles *must* be put on the nose, also another one and a third one, all on top of each other. A razor *compels* them to shave. A comb *must* be used in the hair. A glass *must* be picked up, filled and emptied. Pen and paper that happen to be lying around *must* be used to write down something arbitrary. A turned down bed *compels* a patient to undress, lie down in it and try to sleep. Someone else however, a housewife, began to straighten the bedclothes. Lhermite went all the way. He handed a well-read, distinguished engineer a urinal and, indeed, the engineer passed water in it in Lhermite's presence.

These patients are anything but free; they are slaves of their environment, or rather their brain, they are automatons. Lamme says that we all show this kind of behaviour. All of us will pick up a pen to play with it, or a cigarette lighter that we try out. That is true without any doubt, but we can also stop it as soon as we become conscious of it. And we play with the pen, we don't write with it. We could even call playing the most free form of behaviour, as Friedrich Schiller already argued more than 200 years ago,[42] but evidently not according to Lamme.

Such patients also display imitation behaviour. Smelling a flower, scratching the nose, writing, putting on glasses, everything is imitated. All of that, we also do, contends Lamme. When we fold our hands behind our head, the person we are speaking with will do the same. That is right, we can experience the same when we cross our legs. It even goes much farther. Herd instinct, mass hysteria and the above-described de-individuation can probably be explained in this way.

But are we so completely at the mercy of such things that we have become automatons? We are able, aren't we, probably indeed thanks to a functioning prefrontal cortex, to limit ourselves to a picture of an action without having to execute it? Can't we disconnect ourselves from a situation so that we do not need to copy everyone? Isn't it the ones who suffer from a neurological defect who display automatic behaviour? They apparently follow the suggestions of their mirror neurons without being able to execute 'the free won't'. Lamme denies this and explains how this all happens, according to him.

Hunters' Brain

Lamme demonstrates how an automatic brain functions. When a toad sees a cricket, or a frog sees a fly, and the prey is within reach of their tongue, they will catch it with one flick of the tongue. When the insect is just out of reach a little jump will be required.

In the first case, within reach, the image of the insect appears on the upper side of the retina of the frog. When the prey is out of reach it shows on the lower side. Every part of the retina is connected with a certain part of the frog's brain, the tectum, in such a way that the image of the item within reach has a connection with the tongue and the one out of reach with the hind legs. This is a reflex circuit, so that the frog does not need to go to the trouble of making images or reflecting on things, of which it is not capable anyhow.

Now, if there is an obstacle, this process does not work. The tongue and jump reflexes are then useless. The frog cannot think and therefore does not know how to make a plan to avoid the obstacle, but fortunately it has a reflex circuit for this also. When the entire retina is occupied by the image of an object, such as when the object is large enough to be an obstacle, a different brain structure is switched on, namely the pre-tectum. This sets off another reflex: 'move aside and suppress the other reflexes.' This works because the frog brain is so constructed that tectum and pre-tectum can never both be working at the same time. They mutually suppress each other.[43]

Actually, humans are made in exactly the same way, contends Lamme. In humans the prefrontal cortex has taken over the suppressing function of the pre-tectum, so that we do not need to imitate each other all the time, but in principle this works in the same way as the automatic pilot of the frog. If that were really the whole story, the fact that we allow ourselves to imitate someone – and we have that ability – would testify to great freedom.

The fact that we are indeed fully able to do this makes me suspect that in the course of evolution there has been some essential change.

The Evolution of the Hunt

When we go up a step on the evolutionary ladder we come, for instance, to the snake. When a snake chases a mouse it also uses its eyes (or as in the case of a rattlesnake a warmth sensor). With the eyes it can only see movement. After a bite with the poisonous teeth, the mouse first keeps moving for a while until it collapses and dies. Since there is now no longer any movement, the snake must use its smell to find the dead mouse, even if it is lying right in front of its eyes. To swallow the mouse the snake has to find the head, otherwise the mouse will get stuck in its throat. The snake can only do this with its sense of touch. The snake, therefore, uses three senses separately of each other in the entire operation. These senses have no connection whatsoever with each other. For this reason it cannot make any image for itself.

My cat is a lot better off. It uses signals from different senses in combination: eyes, ears, nose, whiskers and touch in the legs. Those signals together are integrated into an image, which remains in the consciousness of the cat, also when the mouse disappears behind a chest or into a hole in the wall. The cat will then patiently sit there and wait for the mouse to reappear. A snake could never do this. A cat has a working memory with images, and it owes this to its more developed frontal lobe. As a result, the cat has *greater freedom*.

Thus if even a cat is no automaton, why would Lamme want to insist on picturing human beings as automata? The human being as an automaton, as a machine, is not at all a result of modern advanced neuroscientific research; it is an idea with a long history.

CHAPTER 11

The Human Being as a Machine

Every human being is a machine, just like an airplane, only much more complex.

Richard Dawkins[1]

If the body is originally something dead, it can no longer die.

Wunengzi

One of the first experiments performed by medical students in the physiology lab is the electrical stimulus of the muscles in a frog leg, resulting in contraction of these muscles. Not everyone realises – I didn't either at the time – that this is a repetition of an experiment that once led to a huge change in scientific thinking. In 1842, when the industrial revolution was gathering a lot of steam, the German physiologists Emil du Bois-Reymond (1818–96) and Ernst von Brücke (1819–92), devoted many years to these frog experiments, and swore that they would prove the truth of the fact that there are no other forces working in an organism than physical-chemical ones. 'Brücke and I,' Du Bois-Reymond wrote, 'we have sworn to each other to validate the basic truth that in an organism no other forces have any effect than the common physiochemical ones…' Together with Hermann von Helmholtz (1821–94) they were assistants of Johannes Müller, the famous physiologist whose name lives on as the originator of a physiological law, the law of specific nerve energies. It means that every nerve is specialised in a specific kind of perception, no matter what stimulus causes it. Pressure on the eyeball causes light sensations through the optic nerve, a blow on the ear a loud sound via the auditory nerve. About 150 years later, Paul Bach-y-Rita put an end to this idea with his lollypop for the blind.

Müller was an adherent of vitalism, the idea that the life of an organism is caused by 'life force.' As indicated above, in the years preceding the 'sworn agreement' Du Bois-Reymond had occupied himself intensively with the effect of electrical stimulus on the severed leg of a frog. It acted in exactly the same way as a leg of a living frog. He therefore said that it was not 'life force' that moved the muscles, but 'simply' electricity. This had convinced him of the view he passed on to his fellow assistants, namely there is nothing special about life.

Thus this experiment induces in each new generation of medical students implicitly, like 'tacit knowledge,' the same notion, namely that the body is all mechanics. Maybe it is not so much an inducement as it is a confirmation, because the idea that life is a matter of molecular reactions and electrical discharges has been a commonly held view for a long time. Actually, the 'sworn agreement' of these German scientists had been prepared centuries earlier by Descartes who, after cutting into living dogs to study the workings of their hearts, had reached the conclusion that animals were automatons.

Von Helmholtz did not just want an agreement that the only thing that makes organisms function is physics, he also wanted to *demonstrate* it. He did this by proving that the fundamental law of thermodynamics, the law of conservation of energy, also applies to living organisms. This law says that no energy can be lost, but neither can it come into being out of nothing. He did experiments with, again, electrical stimuli on frog legs, of which he measured the generated warmth. However, he did not succeed in reaching the desired result.

He therefore chose an imaginary experiment. He reasoned that, as a consequence of the fundamental law, it is clear that a *perpetuum mobile* machine cannot exist. And because an organism is also a machine, there can exist no 'vital forces' that cannot be traced back to physical causes. He published his conclusion in 1847, five years after the 'sworn agreement,' with the title *Über die Erhaltung der Kraft*. Biologist Rupert Sheldrake, who devoted a chapter of his book *The Science Illusion*[2] to the applicability of this fundamental law to organisms, quoted mathematician Henri Poincaré who said, based on the presumed universal validity of this law: 'It can no longer be verified,' because every outcome that is not in agreement with it can be dispatched as bad science or fraud, or it can be explained by invoking as yet unknown forms of energy.

After Von Helmholtz there have been repeated efforts to show that the fundamental law also applies to organisms, with varying results. In 1980 and 1991 Paul Webb published the outcomes of his experiments in which he found that more energy was consumed than he could explain.[3] When he included the results of earlier experiments by others, he found that the more meticulously the study was made, the clearer was the proof of the presence of energy for which there is no explanation.[4] If we may believe Sheldrake, the applicability of the fundamental law to organisms has not been definitively proven. By the way, the notion that the brain produces consciousness, while this same consciousness then changes the connections in the brain, which again changes consciousness, and so on … looks suspiciously like a *perpetuum mobile*.

Webb's articles did not breach the conviction that organisms are physical machines. Ever since Descartes, and even today, nature only seems comprehensible on the basis of mechanics of human manufacture. In agreement with the quotation at the beginning of this chapter, biologist Richard Dawkins wrote: a monkey is a machine that preserves genes up trees, a fish is a machine that preserves genes in water.[5]

No Spontaneous Generation

In recent times, most (natural) scientists have worked from the premise that material reality is the only thing that exists and that, just as in physics, everything is determined and can be explained in terms of physical cause and effect. In philosophy this conviction is called physicalism, sometimes naturalism, while it is also known as materialism. In this book I have chosen to use the most commonly used term, materialism. This is inextricably bound to the reduction of every field of research to the smallest possible units, called reductionism. *Let us remember, the issue exercising the scientists presented above is about a sworn agreement, not the proven outcome of research.*

Evidently the issue is still not settled. At the funeral of Francis Crick – together with James Watson he had first discovered the structure of DNA, and later he investigated consciousness – his son related that his father was not driven by a desire to become famous, rich or popular, but 'to knock the final nail into the coffin of vitalism.'[6] Apparently that had not yet happened, and he too did not succeed in it. The final nail would have been knocked only if someone had succeeded in creating life from exclusively lifeless parts, something that has been attempted since 1953 but has as yet never been achieved. Craig Venter does make new bacteria and yeasts, but he always uses cells of existing living specimens. Not only has no one ever observed spontaneous generation of life, but paleontologist Richard Fortey even stated a reason for assuming that nature itself had only once succeeded in letting life arise out of nothing.[7] Usually we call such a one-time event in nature a miracle. For all organisms have the same DNA, and that is striking because more forms of DNA are possible that could all equally efficiently perform the coding and transfer of inheritable characteristics.[8] If life had come into being several times, we would have to witness those different forms of DNA in different lineages of heredity. This is not the case. In brief, from the very first beginning life has grown exclusively from other life. Every organism has ancestors, and so far we have not unveiled the *true* secret of life.

11. THE HUMAN BEING AS A MACHINE

No Spontaneous Consciousness Either

Aristotle still believed that a heavy object falls faster than a light one. Galileo did not believe, he measured the speed. Experiments in lieu of notions are the strength of physics. For the early researchers, consciousness did not yet form part of material reality, as proven by Descartes' dualism. Indeed, when we become interested in consciousness, about which the little group of German scientists did not speak, we enter into conflict with materialism. Just as out of physics it is impossible to predict something like consciousness, so it is equally impossible regarding life.

There exists a concept to find a way out of this embarrassment, namely emergent properties. It indicates that in a certain complex configuration of matter something comes about that could not be foreseen. For instance, when a sufficient number of molecules of H_2O come together, the liquid characteristic of water emerges. Thus life and consciousness are described as emergent phenomena of a complex configuration of matter. Because the brain is so complex, we can imagine that it is capable of producing consciousness.

Unfortunately, the introduction of this term does not solve much. In the case of life and consciousness the term emergence is a term of embarrassment that takes the place of explanation which, by the way, does exist in the case of water. There are a number of hypotheses on the origin of life. There are none on that of consciousness. For instance, it is not true that consciousness arises by the presence of nerve tissue. In Chapter 9 above are cited instances of (proto)consciousness of single-cell organisms. These really have no nerve tissue. Just as life does not come about spontaneously, neither does consciousness. Conscious organisms always have conscious parents, and in evolution conscious ancestors.

Materialism stands on the premise that only physical matter exists. That creates problems in the case of consciousness, because of what does consciousness then consist? Particles? Waves? Nature forces? And there is another difficulty: consciousness cannot be reduced to physics without itself first being postulated, for the phenomenon of physics postulates consciousness (of the physicists). Nature itself does not 'know' laws. Natural laws as such only exist in the consciousness of people. As psychiatry professor Thomas Fuchs noted:

> life, consciousness, intention (free will), person (self), all those phenomena that are viewed as illusions, can only be studied because life, consciousness, intention and person exist.[9]

Zombies

Because in the case of consciousness the reduction to matter was clearly not self-evident, it was necessary for the agreement of the nineteenth century German physiologists to be reformulated. Philosopher Daniel Dennett set that as his task and posits in his book *Consciousness Explained*:

> The prevailing wisdom, variously expressed and argued for, is materialism: there is only one sort of stuff, namely matter – the physical stuff of physics, chemistry, and physiology – and the mind is somehow nothing but a physical phenomenon.[10]

'The prevailing wisdom': Dennett is wise enough not to speak of proven wisdom. The consequences include for him:

> Are zombies possible? They're not just possible, they're actual. We're all zombies. Nobody is conscious — not in the systematically mysterious way that supports such doctrines as epiphenomenalism. It would be an act of desperate intellectual dishonesty to quote this assertion out of context![11]

As he had previously explained, in his use of words, zombies are identical to automatons. And as I have indicated in the foreword, these are not conscious.

11. THE HUMAN BEING AS A MACHINE

Consciousness as Epiphenomenon
The epiphenomenalism in the quotation from Dennett is one of the first materialistic solutions that was formulated for the brain-versus-consciousness problem. The problem is the following: how can something non-material like consciousness have an influence on the material brain? Because in the materialistic conception this is not possible, Thomas Huxley (also known as Darwin's Bulldog because he was Darwin's fiercest defender) in 1874 embraced the view that consciousness is produced by the brain, but is not capable of taking initiative or being the cause of anything in that brain. This is called epiphenomenalism. Huxley compared consciousness with the whistle of a locomotive; it does not contribute a foot to the movement.

Evidently, Dennett has no interest in epiphenomenalism. If consciousness had no role in what we do, this would be a denial of human culture. Therefore there had to be another way out that also would not be incompatible with the mechanistic world view. Consciousness would then have to be physical, simply an aspect of the physical brain, as Dennett formulated it above. If that is so, says philosopher Thomas Nagel in his book *Mind and Cosmos*,[12] it would mean that consciousness is, as potential, inherently present in physical matter. That means: from the Big Bang. Consciousness would then have existed ever since the beginning of the universe. If so, intelligent design, or theism as Nagel calls it, is not far away. That cannot have been the intention of materialism, can it? For materialism not only wanted to do away with vitalism, but also with 'the heavenly foot in the door'.

Consciousness as a Thing
Consciousness would therefore be a thing: the brain. We have already rejected this idea on neuroscientific grounds, with the agreement of neurophilosopher Libet (see Chapter 6), but here it is introduced as a philosophical idea. The brain, the fatty organ in our skull that weighs some 1300 grams, and consciousness are identical. This is called the psycho-physical identity theory: $\Psi = \Phi$ (Psi = phi).

143

It would of course mean that consciousness resides in the skull. The shared identity of such different categories is compared by its defenders with *water = H$_2$O* or *warmth = molecular movement*.

If you now get a feeling that something is askew here, that this offers us nothing like an explanation of the nature of consciousness, and that our feelings and experiences cannot be measured or weighed, and are therefore not material processes, you are right. Scientific research has proven that water is H$_2$O, and the same is true for the relationship between warmth and molecular movement. This is not true for brain = consciousness. Consciousness is not *the* characteristic of brain matter, not even of brain processes, but it is a new, unforeseen, if you will, emergent property. Therefore it is not the same thing as the physical brain.

One of the seemingly most convincing arguments for the identity theory is the fact that there are medications that can change consciousness. What these actually do is to connect themselves with receptors in the brain, which results in a change in the number of neurotransmitters and neuromodulators. This change correlates with the disappearance of feelings of melancholy or fear. The medications themselves have no feelings, so we should call antidepressants as they are known scientifically: selective 'serotonin-reuptake-inhibitors' and tranquillisers 'selective gaba enhancers.' For whether these medications will work is entirely uncertain.

Feelings depend on many more factors that just the chemicals in the brain, factors that relate to the person himself or herself, their history and environment. What is interesting is that in cases of depression a placebo has approximately the same effect on the brain and causes virtually the same change in feelings as medications do.[13] And in both cases it takes the same amount of time before they take effect, namely the same time it would take spontaneous improvement to occur.[14] The latter finding could mean that the organism is delivering an autonomous achievement, animated by the therapy. Prescribing medications is not an administrative action, but an action between two people. Thus counseling also has similar effects on the brain.

Consciousness as Secretion

Even some neurodeterminists continue to have trouble with the consequences of the view that the brain and consciousness are identical. 'Just as the kidneys produce urine, the brain produces consciousness' is a favourite one-liner of Dick Swaab.[15] And he is neither the first nor the only one to make this comparison. René Kahn, professor of biological psychiatry, said:

> The brain is an organ like any other, such as the pancreas which produces, among other things, the hormone insulin. ... The brain produces in addition to hormones an even more important thing: it produces behaviour.[16]

Kahn is thus a little more careful; he does not mention consciousness, but behaviour as the brain's product. And as we will see later, that is a good idea. But also in that case, it is not even one half of the story, for behaviour does not exist without a body and without a context.

Variations on this theme have been around for over two centuries. It started with French physician Pierre Cabanis, who said: 'The brain secretes thoughts like the liver secretes gall.'[17] But just like his French predecessor, Swaab fails to explain how brain processes become experiences. So far, no one has discovered this. According to neuroscientist Susan Greenfield, the biblical riddle of the change of water into wine is easier to solve. Consciousness philosopher David Chalmers therefore called it the hard problem of consciousness. Kahn's preferred solution is derived from the psychological school of behaviourism. In this view, the mind, consciousness, is utterly unimportant; the one and only thing that counts is behaviour. The transition from brain processes to behaviour is not an equally hard problem. A true neurodeterminist almost has to be a behaviourist.

But what is really expressed here is that brain and consciousness are evidently not identical after all. The one is supposed to be a product of the other. This raises new problems. For just as we cannot compare the relationship of brain and consciousness with that of H_2O and water, we also cannot compare consciousness with gall, urine or insulin. As opposed to gall, urine or insulin, consciousness cannot be described in materialistic terms. For what is it made of? This is the reason why experiences and feelings – in brief, consciousness – cannot be considered real in the materialistic model. In most 'brain books' consciousness therefore continues to be considered as an unimportant epiphenomenon, a side-effect of the genuine work of the brain, without significance, a chatterbox or even an illusion. And, importantly, it is therefore not capable of bringing about anything in the material world. The brain does that completely outside of consciousness. This is the reason, for instance, that free will cannot exist. Now, an illusion presupposes consciousness, and therefore we cannot explain consciousness away in this manner.

Energetic Consciousness?

And here is another point. Suppose that consciousness is produced by the brain. In the meantime it has become clear that consciousness also causes changes in that same brain so that, if these were both true, the result would look, as already said, very similar to a *perpetuum mobile*.

Now it becomes interesting to take a look at the relationship between consciousness and energy. In the previous chapter we have seen that when we have to exert our brain (with self-control, conscious system 2) we can quickly become exhausted, a problem that can be solved by eating sugars. This kind of exertion therefore suggests a high level of energy use. But why does conscious thinking take so much energy? Is it because the brain has to work harder? Why do brain processes that occur automatically take less effort than conscious processes? Is it because thinking causes new connections? That sounds credible, for this requires the delivery of proteins along the entire length of the nerve fibres.

But what about sleep? That takes no energy at all. Of course not, one might presume, for at night the brain rests. But this is not the case. We may sleep, but our brains are on night shift, which is even busier than its activity during the day.[18] The same areas that were used by day are activated once again, but now without corresponding consciousness, and six times as fast! How is it possible that this does not make us tired? The reason for this activity appears to have something to do with the consolidation of memory by strengthening connections and making space available for new experiences. But does this activity indeed take less energy?

One research project seems to offer some insight into this riddle. It turns out that in the first hours of sleep, during deep sleep, a great 'wave' arises of the substance that carries energy in our organism, ATP (adenosine triphosphate).[19] This means that suddenly there is a large quantity of potential energy present. But what is the significance of this? The researchers assume that this ATP is made to effect restorative measures in the brain, in the form of the production of proteins. In the journal *Sleep* Margaret Wong-Riley wonders whether this is right. Protein production does not require that much energy. Most of the energy in the brain, she contends, is consumed to make electrical discharges in the neurons possible. In other words, to restore neurons after a discharge. However, the amount of discharges does not differ much between day and night.

But why would neurons make ATP they do not use? She reasons that it cannot be extra production, but an unused excess of ATP that evidently is mobilised by day and not by night.[20] But what for? She doesn't say. What else can it be for than daytime consciousness? But why does that take energy? Because 'producing' consciousness takes energy? But then energy disappears, and that is not allowed by the law of the conservation of energy. Or is it because we only think when we are awake, and would this especially create new connections? (It seems to me that this is indeed allowed by the law of conservation of energy.)

If consciousness can indeed be the cause of things, for instance the growth of new connections, it seems to me obvious that the energy required for making these connections simply comes from the mitochondria of the neurons. Thinking is what most of all distinguishes *human* consciousness from that of so-called 'non-human' animals.

How important this mitochondrial energy is precisely for human consciousness is demonstrated by the following.

Biologist Lawrence Grossmann wondered how the brain cells, which have but a small cell body and consist primarily of thin nerve fibres, are able to satisfy their energy requirements. A cell needs mitochondria for this, little cell organs which however, are too big to fit in large numbers in a nerve cell, let alone in the thin fibres. These mitochondria produce ATP. The brain consumes the lion's share of the available energy; its weight is only 2% of the total body weight, but it consumes 20% of the energy when in rest. It has been shown that since the transition of Homo erectus to Homo sapiens, some 58,000 years ago, eleven mutations have improved the energy management of these mitochondria. That is a great deal, especially when compared to the 25 million previous years, which generated exactly one mutation. 'We were surprised when we saw such big changes happening in a short evolutionary period.'[21]

And yet, this is not the solution of the conflict with the fundamental law. For how can non-material consciousness cause something to happen in the physical world? That implies that we should also be able to move a glass of water with thought power. According to the fundamental law that is impossible. But all things considered, we do exactly that all day long. Whenever we think of picking up a glass of water, we set our muscles in motion to execute the action. Does that mean that our daily experience is an illusion? After all, the idea that consciousness steers our actions cannot be accepted by natural science. The neurodeterministic solution of this continues to be almost identical to epiphenomalism, namely that our brain produces both the thought and the deed of picking up the glass, more or less at the same time. Without thought, without consciousness therefore, it would be possible just as well. Consciousness has no power. We are zombies.

Thought Power, Really?

That is the materialistic model. But now, let's look at the real world. Of course, consciousness exerts influence on the brain. How else

11. THE HUMAN BEING AS A MACHINE

can we explain that thoughts and experiences cause changes in nerve connections? How can non-material consciousness change those connections? We have seen that the answer 'the brain changes itself,' which indeed fits in the materialistic model, does not hold true. See the story of Pedro Bach-y-Rita in Chapter 2.

Moreover, for a long time already people have made use of the causal difference made by consciousness in the functioning of brain cells. Dick Swaab relates on the last page of his book *We Are Our Brain* how electrodes were implanted in the brain cortex of a patient with paraplegia, who was then able to steer a computer mouse and subsequently an arm prosthesis by thought power. (I suspect that he means the little arrow on the screen.)

This specialty is experiencing enormous development, in part in the form of games in which we, wearing a little crown with electrodes and merely by concentrating, can write our name on a screen and even drive a car. The concentration appears to be of rather short duration, for it takes a fair amount of will power (= energy). Researchers are now working on methods to communicate with 'locked-in' patients, completely paralysed patients who are conscious but because of their paralysis cannot communicate. Consciousness is therefore the only faculty these patients still have at their disposal – consciousness without behaviour, a nightmare for behaviourists! The patients learn to point to letters merely by thinking thoughts of which, although they have no relationship to the letters, the connection with specific letters has been arranged beforehand.[22] For that matter, the thoughts are always visualisations of actions or movements – such as movements of the hand or tongue; they are never word-thoughts. Thoughts can therefore not be directly read.

In practice, neurologists apparently completely ignore the theoretical objections to the active role of consciousness. Of course, this does not prove that consciousness itself possesses physical energy. The energy needed for brain processes, just as for all other tissue and organs in the body, comes from mitochondria in the cells, which consume glucose to generate it. It is an important reason why it is helpful to eat some carbohydrates such as a cookie, before we devote ourselves to serious thinking.

Consciousness Is Not Physical

In Chapter 6 we discussed an experiment by Benjamin Libet in which a stimulus applied directly to the sensory brain cortex became conscious after a later stimulus applied to the skin. This led him to the conclusion that the identity theory is incorrect, and that consciousness is a non-physical phenomenon. What then is the reason for the trend in science to insist on viewing organisms as machines and consciousness as physical? Was it a coincidence that the Germans' 'sworn agreement' to spread this idea occurred during the Industrial Revolution? It is apparent that the answer has to be sought in the collaboration, or rather the lack thereof, between the two hemispheres of the brain.

CHAPTER 12

Two Hemispheres Under One Roof

> *Soccer players are creative people. They generally use their right brain. At school, when we are learning, the left side is filled. When the left starts to dominate the right, collisions result.*
>
> Johan Cruyff (renowned Dutch soccer star.)

Why do our brains actually have two hemispheres? One as a reserve? Alas, no. An injury to one of the two hemispheres almost always leads to some handicap which, by the way, in some cases can be overcome, as we have seen in several examples. We can also widen the question. Why do we have two eyes, nostrils, ears, lungs, nipples, kidneys, ovaries, testicles, arms and legs? In the evolution of the animals symmetry developed quite soon, notably in those species that also developed a central nervous system, from squid to whale, from mosquito to elephant. And, notably in the vertebrates, these two sides are each served by their own brain hemisphere. But that does not explain why both hemispheres have such different tasks.

Both hemispheres are connected by a wide flat bundle of nerves, the *corpus callosum*. We might think that this connection would enable the two hemispheres to work nicely together. To a certain extent that is indeed true, but most of these tissues have the task to suppress the other half when the one is at work. What can be the reason for this? It would only make sense if the different hemispheres had conflicting tasks. This indeed seems to be the case.

And it goes much farther than the idea that the left brain is straight-line-logical and the right brain creative. Or the contrast between language in the left brain and orientation in space in the right brain. For why would that create a conflict?

To make things even more complicated, we know, on the other hand, that both hemispheres are certainly occupied with the same things. Only, they each do that in a different way. Each provides, as it were, its own form of attention. This holds true for all vertebrates. In order to understand why both hemispheres each give us their own view of the world we have to go back in evolution.

A Chicken's Eye for Details

Most vertebrates have their eyes on the sides of their heads. To stay alive, they have to find food and not become food. For the first purpose they have to be able to focus and recognise food; for the second they must not focus at all, but keep an eye on their whole environment. These are two conflicting forms of attention the eyes must have for the world. Each of these two is therefore entrusted to its own eye. The one seeks details, food, that which it knows; and the other watches the surroundings, the big picture, alert for anything unexpected. Animals with eyes on the sides of the head see the left field of vision with the left eye and the right field with the right eye. These are each projected contra-laterally into the visual brain cortex, the left field in the right hemisphere and the right field in the left.

Just watch how a chicken scratches in the ground while it looks around rather absentmindedly. Then she makes two steps back, turns its head askew to inspect the ground with its right eye (left brain) for anything it recognises as food. In the meantime, the left eye (right brain) looks into the sky to see if there is perhaps a hawk or some other danger. After all, potentially threatening animals always appear to the small chicken from above its own horizon. Also in the enclosure, when there is no danger from above, they first peck things on their right and then on the left, while they watch me with the left eye – expecting something? Or suspiciously? If there is something on the ground the chicken does not recognise, it will turn its head and take another look with the left eye.

In *The Master and His Emissary*, a study of the right-left differences of the brain, psychiatrist Iain McGilchrist reviews a whole series of animals that show the same division of interest of the left and right eye: ravens, cats, chimpanzees, toads, magpies, avocets, rats, frogs and crows. The right eye sees detail, the left the whole. The right eye concentrates on what is known, the left on the unexpected. Several of these animals, by the way, have their eyes in front. Nevertheless, due to the partial crossing of the visual nerves, their visual system is so organised that they process the right visual field in the left brain and left field in the right one. Predators and birds of prey that have their eyes in front use their right eye and leg to strike or catch a prey.

Gouldian finches choose their partners with their right eye. Some of them have red heads, others black heads. Red males preferred red heads, black males black ones. When the right eye was taped shut they became confused and showed no preference. Thus they choose with the left brain based on a familiar detail.[1] Evidently, they are only looking for detail, which is processed in the left brain. In humans, at any rate in men, this may also be true for generally attractive female body parts, but love is experienced in the other hemisphere. For humans, faces are important, and these are handled in the right brain. When test subjects, lying in an fMRI scanner, were asked to think of their loved one, they showed the activity that was deemed to correspond with this only in the right brain.[2]

Asymmetrical Symmetry

There are of course individual differences, just as not all people are right-handed. In the West around 89% of people are right-handed, and most of these have their speech and semantic areas in the left hemisphere. This will hereafter be our standard when we speak about the left and right hemispheres. The remaining 11%, the left-handed ones, in part follow the standard pattern with regard to the speech centres on the left; thus only 5% of people in the West do not have their speech centres in the left hemisphere. In some of these everything is simply reversed: everything that is on the left in the standard pattern is on the right in them, and the reverse. And there is a small group in whom we see a partial reversal, which often involves diverse other abnormalities such as schizophrenia, dyslexia, some forms of autism and of *savant* syndrome (people, mostly with mental disabilities, who have extraordinary skills, especially in memorising, or in calculating, or sometimes in arts, skills that 'normal' people don't have).

But if we stay with the standard we see the following picture. In social mammals, and also in humans, the right hemisphere, especially in front, is larger and more voluminous than the left. Only on the side and in back (parietal lobe and occipital lobe) is the left hemisphere larger than the right one. The hemispheres are therefore asymmetrical and look as if rotated a little. This is called the Yakovlev torque.

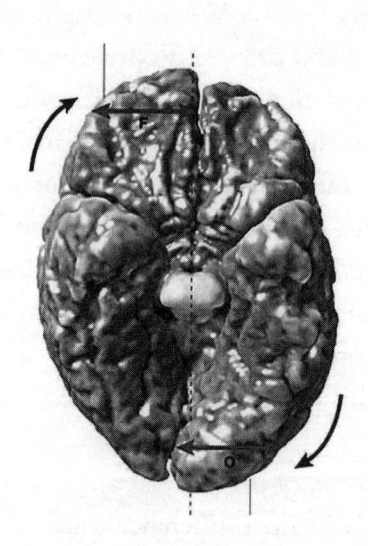

Figure 10. The brain from below

The right brain (left in the picture) is a little bulkier; it has more and bulkier nerve cells with more branches and connections than the left, and also has more white substance. This all indicates more global and faster connections, while the left has more local connections.[3]

The right frontal lobe is significantly larger than the left, while on the right the parietal and occipital lobes have less volume. On the left the occipital lobe protrudes a little. Thus there is a larger impression in the skull in the right front and left rear. These are called the frontal and occipital petalia. These are also found in apes and in Homo heidelbergensis of some 400,000 years ago, the probable ancestor of both Homo sapiens and Neanderthals.

Denial

We are evidently witnessing two different ways to view the world. To characterise these I will discuss them as if they were living persons, despite the fact that I have been fulminating against that. The reason is readability; it makes the discussion shorter and more lively, and I would like to keep your attention. Nevertheless, of course, I do consider both hemispheres of the brain as instruments.

As we have seen, the right hemisphere serves the left half of the body and the left hemisphere the right half. But the left brain has a one-sided understanding of its task as compared with the right brain. Just as the right brain is concerned in a global manner with the whole world, it acts in the same way on the entire body. Not so for the left brain; this limits itself to the right half of the body.

Neglect

This limitation of the left brain has interesting and characteristic consequences for the symptoms that appear in case of failure of one of the two hemispheres. In the previous chapter we learned about an association area that, according to V.S. Ramachandran, is typical for humans. It is the curved ridge around the end of the Sylvian fissure, the gap between the temporal lobe and the rest of the brain. This ridge is called the inferior parietal lobule, IPL (see Figure B, inside front cover). Although it is already present in lower mammals, it grows in primates. In the human being it has grown to such a size that science distinguishes two parts in it, the *gyrus angularis* (GA) and the *gyrus supramarginalis* (GM).

These are situated on the crossing of the occipital (vision), temporal (hearing), and parietal (body in space) lobe, and the association area of touch (see Figure B on inside front cover).

When in a person the IPL of the left hemisphere is damaged, he shows *ideomotor apraxia*, meaning that he is unable to transform an idea of an action into an actual action. The awkwardness manifests primarily in the right half of the body, the part related to the left brain. This means that when told to make a certain movement with the right hand, he is incapable of executing it. If asked to do as if he were combing his hair, he will lift his arm, look at it and swish it around his head. When asked to do as if he were hammering a nail into the table, his response will be to strike the table with his fist instead of gripping an imaginary hammer. In response to the request to make a military salute he will helplessly look at his arm and do nothing with it. In brief, the person cannot create a mental picture that precedes an action. He cannot recognise objects by touching them with the right hand.

However, when the same area in the right brain has been disabled but the left side is healthy, something very different appears, namely the remarkable symptom of *neglect*. The entire left half of the body is ignored; its very existence is denied! A patient with neglect will say of his paralysed left leg that it is not his and that someone must have put it in his bed. It is as if the still functioning left IPL is only interested in the right half of the body, while when the right IPL remains intact, this is still willing to take care of the whole body. In both cases something is wrong with the *body schema*, which is the term for our awareness of where all our body parts are, spatial body awareness. *The left brain is therefore interested exclusively in the right half of the body; the right brain in the entire body.*

In case of further damage to the parietal lobe, so that the visual (association) cortex is also affected, the patient can no longer see the left vision field, and will consistently deny that anything is wrong. Figure 11 shows how a neglect patient copied the pictures on the left. In reverse, when the damage is in the left hemisphere and the right one is intact, this does not happen. The left hemisphere actually causes a narrowed view of the world.

Figure 11. The drawings on the right are attempts of a neglect patient to copy those on the left

Built-in Ambivalence

McGilchrist speaks of two portals of consciousness. We could say, the one has its eyes wide open, the other narrowed. He too apologises for treating the two brain hemispheres as two personalities. They represent two different kinds of consciousness, he says, for when we speak of personalities, we also include people's consciousness. I would actually like to view them as two different forms of intelligence that each create their own picture of the world, since intelligence is a quality of the brain that is at least in part acquired (see Chapter 2). Each of these forms of intelligence has its own influence on consciousness, culture and science.

Does the Brain then Determine Everything After All?
But that would mean that this evolutionary inheritance determines our consciousness. Are we then our brain anyhow? McGilchrist says he does not believe that the brain produces consciousness:

> Is consciousness a product of the brain? The only certainty here is that anyone who thinks they can answer this question with certainty has to be wrong ... the one thing we do know for certain is that everything we know of the brain is a product of consciousness. That is, scientifically speaking, far more certain than that consciousness itself is a product of the brain.[4]

But still, his reasoning certainly seems to indicate that it makes a difference which hemisphere has primary control of consciousness. How does this work? Do the hemispheres determine consciousness, or has this split situation come into being *as a consequence* of the influence of consciousness, as I have argued so far in regard to the position of the brain? Actually, both are true. The right-left difference has its roots in evolution. We cannot change that, just as we cannot change out gender (other than by operation), our country of birth and the colour of our hair (other than by dyeing). And yet, we begin life with a larger right brain, which is the case for two years, and after that the right brain remains dominant for another year. This is understandable because the first thing a child does is orient itself in the world. Everything is new. The child crawls and looks, and exercises its visual-spatial capacities, which is reflected in neural processing in the larger space available in the right brain cortex.[5]

Brain scans have shown that a mother also uses her right brain to communicate with her baby. She constantly shows the child that she perceives its emotions. She names them in a meaningful tone, shows matching facial expressions, and reassures and cuddles the child. This gives the child confidence in the mother and in the world.[6]

12. TWO HEMISPHERES UNDER ONE ROOF

In those first two years, the child still uses both hemispheres to process sounds and also the words of the mother and other caretakers. Then the child begins to speak – still to orient itself – so that it can learn in even greater detail from the experience of the caretakers, and for this it uses primarily the space in the left brain. Norman Doidge says in his book *The Brain That Changes Itself*:

> In other words, each hemisphere tends to specialise in certain functions but is not hardwired to do so. The age at which we learn a mental skill strongly influences the area in which it gets processed.[7]

Thus we see that the use we make of the brain definitely determines the way it functions.

When in a young child one of the hemispheres develops a problem, the other side still easily takes over. Probably this is possible because small children have 50% more nerve connections than adults. In normal cases the unused ones are 'pruned back' around the time of puberty – 'use it or lose it' – probably already around the ninth year or even earlier.

Mars and Venus in the Brain

Gender differences have been found in the two hemispheres of the brain.[8] Men have more connections in each hemisphere separately, especially from front to back and also in the cerebellum. Women have more connections between the two hemispheres. This means: in general. In individual cases one cannot speak of a masculine or female brain. The researchers suggest that this indicates a better connection between perception and coordinated action in men in general, and better communication between analytical and intuitive ways of thinking in women. Men and women thus complement each other. It is shown that this difference does not exist at birth but grows only after twelve to fourteen years, around the time of puberty therefore. This will not surprise us, because at that age sexual maturity occurs, with the role assignment that seems to form part of it. In view of the fact that culture affects our brain connections (see Chapter 9), it is possible that this phenomenon is not only hormonally caused but also by expectation patterns of what is considered masculine and feminine.

Plasticity in Older People

But that is not the complete picture. As we age, a new role assignment grows between the two hemispheres of the brain. In principle, in the absence of brain illnesses, the life of the brain cells simply continues until death. An autopsy of the brain of Hendrikje van Andel, who died at age 115, showed that her brain had remained completely intact. What does take place is a decline in volume after our fiftieth year, and even more pronounced after our eightieth.

Apparently, healthy seniors may react to this with their own kind of plasticity. They mobilise greater quantities of neurons in the affected areas by overcoming the relatively strict separation between the tasks of the left and right hemispheres; in other words, the lateralisation of brain activity is diminished, especially in the frontal lobe. This is known as the HAROLD model (hemispheric asymmetry reduction in older adults). Besides the usual areas, older people thus also use areas in the other hemisphere for their tasks.

As people age, therefore, the strict separation in the tasks of the two hemispheres decreases, and both forms of intelligence develop more balance. That might well be the reason why wisdom comes with the years. I am always struck by the fact that valedictory addresses of professors, who used to display a fighting spirit and fanatically defend a particular point of view – which mostly points to dominance by the left brain – suddenly sound much milder. And some power-hungry politicians become champions of human rights and sustainability after their political career.

12. TWO HEMISPHERES UNDER ONE ROOF

But before we get there, we are stuck with the situation sketched by Johan Cruyff: 'When the left starts to dominate the right, collisions result.' Both hemispheres give us their own view of the same situation, one could say. The one can suppress the other. What happens then is the subject of the next chapter.

CHAPTER 13

Conflict Between Neighbours?

In school they were written on the blackboard,
The verb to have and the verb to be;
This indicated time, eternity,
The one reality, the other appearance.

Ed. Hoornik[1]

There is a brain scientist who can give a good explanation of the differences in consciousness that serve the two brain hemispheres: Jill Bolte Taylor. She is not only able to do this because she has researched it so well at the University of Indiana, where she teaches, but primarily because her own left brain was out of commission for a considerable time due to a brain haemorrhage.

Blood is toxic to brain tissue. That is why the cells in the blood vessel walls are packed close together so that no small particles, bacteria or hydrophilic molecules can seep through, the so-called blood-brain barrier. During the four hour duration of her growing haemorrhage, Jill Bolte Taylor, due to her expertise, was able to follow exactly to what extent the failure had progressed. She noticed when it became clear to her that she had to call for help, that she could no longer verbalise thoughts and telephone numbers. At the same time she noticed that her inner dialogue stopped. Bolte Taylor described what she subsequently experienced as a feeling of being enveloped in an all-encompassing unconditional love and unity with the universe ('la-la-land') without past or future.[2]

13. CONFLICT BETWEEN NEIGHBOURS?

Jill Bolte Taylor has completely recovered. The process took eight years, during which time she had to exercise everything without any help from the motor cortex or the speech cortex, which she had lost. Just like Pedro Bach-y-Rita, she had to get herself to the point that other areas became able to take on those tasks. It is perfectly clear that here the function preceded the form, the structure of the brain. Ten years later she had come to the point that she could write about it. She now possesses again all functions that originally belonged to the left brain. And how! Her speech is like a waterfall, as I experienced in her presentation of the Dutch translation of her book.

Thus we can see how it is possible for a person, by hard practice – in her case with the help of her mother – to restore the brain. Her mother had the same intuition as the sons of Pedro Bach-y-Rita: practise from the level at which children begin. The first thing the mother did was to climb into bed with her paralysed daughter, who had also lost her speech. Otherwise, the recovery turned out to be a mixed blessing, because now Bolte Taylor knows again what negativity is: how she can judge people much too critically, including herself. Since she now knows that the left brain plays a role in this, she has taught herself to focus on her right brain. How does she do that?

She does it by silencing her inner dialogue, in which also critical judgments and other forms of negativity are played out. The means is meditation, inner silence or concentration on a thought, a sentence, a sound or a picture. Then she succeeds again in achieving a state of 'union with the universe.' She thinks that apparently it is language that blocks access to the right hemisphere – we inwardly talk and interrupt too much. She explains that the right brain is a parallel processor who can follow many tracks at the same time, whereas the left brain is a serial thinker. The right brain lives in the here and now, in cohesion with the whole, and the left brain lives in the past, that which is familiar, and in the details, and it projects those into the future. In her view, the two hemispheres therefore also do different things with incoming information – two forms of intelligence, that process information differently and generate a different picture of reality.

Detail and Context

Each of the two hemispheres has its own perception capacity, and they complement each other. Both are indispensable. In the meantime, an abundance of neurological cases in which one hemisphere had failed in whole or in part, has demonstrated the differences by fMRI and EEG tests,[3] namely that the left brain can only handle what is familiar and cannot use information that does not agree with what it already knows, while the right brain is delighted with surprises. The right brain perceives and understands things in context, the left brain can only form abstractions, which means separating things from their context. 'The patient declares that it is winter because it is January, not by looking at the trees.'[4]

This also indicates that the right brain is interested in the environment, nature and landscape, whereas the left brain especially feels at home with man-made things, things that can be understood, manipulated and owned. It is the reason why the left brain can understand nature only as a mechanism. The left brain is only interested in the literal meaning of language and does not understand deeper layers of significance, no original metaphors, no humor, irony or sarcasm.[5] Sudden insight in the point of a joke or the solution of a riddle, when one has to think out of the box, is accompanied by sudden activity in the frontal superior temporal gyrus of the *right* hemisphere (see Figure B on the inside front cover).[6] According to McGilchrist: a sentence like 'it is hot here today' leads someone who possesses an intact right hemisphere, and can therefore think out of the box, to opening a window. If due to damage to the right brain only the left side can get involved with this, the sentence will be accepted as a meteorological statement.

The left brain cannot distinguish a joke from a lie; it only notices the 'what,' whereas the right brain is also interested in 'how.' Left: quantity, right: quality; left: to have, right: to be. The left side likes to categorise and put new things into known categories (indicated by phrases such as 'nothing but' as in 'genetic modification is nothing but selection; people who improve seeds have done it for ages.') The right side sees the individual 'Gestalt' of something. Facial recognition and recognition of emotions in faces happen in the right hemisphere. The left brain – remember the use of tools by the crow – is most interested in tools, instruments and machines, man-made things; the right brain more in what our (natural) surroundings bring to us.[7]

Science

The sworn agreement made in the nineteenth century by Du Bois-Reymond and Brücke, implying they viewed a living organism as nothing but a lifeless mechanism, now becomes understandable. It seems obvious that natural science has to rely on the left hemisphere. But science is about knowledge of the world, and both hemispheres can contribute to this. They each give their own particular direction to this knowledge.

When we say that we know someone we possess a knowledge that cannot be conveyed to someone who does not know the person. No words suffice to do that. All we can do is formulate facts, such as length: 5 feet 10 inches, dark brown, medium-length hair, brown eyes, nice figure, likes classical music, traveling, photography and gardening, not a bar type. She looks for broad shoulders to lean on. Yet, to really know what she is like we will first have to meet her. The wealth of information a meeting will generate was called by physicist-chemist and philosopher Michael Polanyi tacit knowledge – silent, implicit knowledge. It is the domain of the right brain.

The left brain only gives access to facts that can be formulated; for the rest, it has to consult the right brain. Science only allows access to explicit knowledge, in words, figures or algorithms and pictures. Implicit, tacit knowledge, which also exists but cannot be communicated verbally, can never find a place in a scientific article. But that does not mean that it does not play a very large role. The right brain plays a role in conveying customs, the generally accepted but in part non-formulated views of what is scientific and what is not. And it probably makes the ideas possible that can lead to research.

Natural science made explicit depends for its explanations on the left hemisphere. We could call implicit knowledge *living* and explicit knowledge *dead*, for the latter has become crystallised and definitive. It is like the difference between characterising and defining something. Just try to define your loved one, or a piece of music. When that is attempted, when for example we try to define love – whether it is love for human being or for art – neuroscience will produce extremely interesting, undoubtedly true but totally dead, facts that completely lack the warm and living nature of love in its contextual and relational reality. Love and feeling for beauty are then the results of pheromones, dopamine, oxytocin, and of a Darwinian preference for symmetry in humans and animals, and a specific waist-to-hips ratio in women that is supposed to indicate 'good genes.' It does not go beyond clumsy attempts when we want to investigate the neurophysiological reasons of the experience of beauty and love, the way, for instance, Ramachandran does in his book *The Tell-Tale Brain*.

The Left Hemisphere and the Natural Sciences

As already cited, psychiatrist Iain McGilchrist has written on the influence of the two hemispheres on culture.[8] He took the title of his masterwork *The Master and the Emissary* from a story by Nietzsche about a king who sent his emissary out to rule the country following the directives of the king. Because in the opinion of the emissary he does the actual work, he undertakes a coup and seizes power. McGilchrist compares the king with the right brain and the emissary with the left.

The coup by the left brain, according to McGilchrist, is visible in western culture, and the resulting influence of this culture on the brain again generates a preponderance of the left brain. Due to the fact that the left brain is able to suppress the right brain we are witnessing a self-reinforcing process. The book is a critical review of cultural history, in which the author shows that this coup has occurred repeatedly, especially in western civilisation.

13. CONFLICT BETWEEN NEIGHBOURS?

But the world – in the form of the natural environment and the human body – always resists the cerebral coup of the left brain. This was successful until the industrial revolution drastically changed the inhabited, and even the uninhabited landscape. Both cities and agrarian landscapes were given utilitarian forms. Straight lines and right angles, which occur nowhere in nature, determined the environment from then on. And the environment determines consciousness to an important extent. We saw that in the Müller-Leyer illusion, which primitive peoples do not experience.

The Western world today constantly confirms that the left brain is 'right.' The right brain has to function without such 'natural' nourishment. Add to this the fact that the brain is constructed in such a way that what we use becomes stronger, and what is neglected starves and withers. It is a case of winner takes all, for the active hemisphere suppresses the other more and more. This is the reason why the left hemisphere has never before shown such strong dominance as in our time. No wonder the sworn agreement of the three German physiologists – about the materialistic and mechanistic character of living nature – occurred during the Industrial Revolution. It is also the reason why no one sounds the alarm when we hear that our body is built up from cells, just like a machine is built from parts. But that is of course not true, for we began as one single cell and ever since fructification began we have been *subdivided* into cells, not *built up*. We are one whole. The right hemisphere sees that right away with 'common sense.' The right brain is holistic.

The idea that we have an image in our head could only arise after the discovery of optics by Euclid and its further development – with the camera obscura and lenses – in the time of Descartes. Man-made apparatus provides us with metaphors that are easier to understand for the left brain than what is offered by daily experience, through which we see the image 'where it is.'

It is also the reason why we immediately embrace the idea that the brain makes consciousness. I can hardly escape it myself. It seems to agree with our insight in the machinery with which we surround ourselves.

But it is also an insight the consequence of which – 'we are our brain' – makes many of us feel uneasy, and this is something the right brain is responsible for. Two or three centuries ago virtually no one would have embraced this idea.

It Works in Practice? Fine, Does It Also Work in Theory?

Because the left hemisphere thinks without the context of the world, it can slavishly follow the inner logic of a theory, even if it moves farther and farther away from everything that experience teaches us. These very sorts of theories possess a certain intellectual attraction, such as the idea that we construct an illusory picture of the world for ourselves instead of simply observing what presents itself. Or the conviction that we feel pain in our calf not where we feel it but in our brain. Or the notion that we do not control what we do or think. The right brain is occupied with what is present, and the left brain with the signs that point to present reality, such as numbers. Quantifying things at the expense of evaluating quality is a practice that has spread over all the sciences by now. Not until something can be computed is it convincing. In my profession, for instance, a therapy only has value if a category of patients reacts favorably to it, not if someone benefits from it. No, not even then; only if in a category of patients a target value is reached which, according to theory, has to be as much as possible above or under a specific number, also if it makes someone not feel well. In the meantime it has been shown that patients for whom such target numbers were reached (blood pressure, cholesterol, or blood sugar in case of diabetes) were worse off than those who had been treated less strictly.[9]

In natural scientific research mathematical, read computer, models are used. Mathematical models demand far-reaching abstraction and simplification. That can actually only be done with validity in the case of man-made machines and other objects. But model making does not stop there. The widely observed weather forecast already shows that is not so simple to make a model that is true to reality. The same holds for the neuroscientific model of the relationship between the brain and consciousness. In this way it is not difficult to see how ideas like 'we are our brain, therefore we cannot change after adolescence' find acceptance; also notions like 'free will does not exist' and 'we are automatons' or 'consciousness is a chatterbox,' even if someone *voluntarily* writes a book to convince readers of these truths. The left brain sees things where the right brain sees beings. The left brain reifies, turns life into a thing. We owe the mechanisation of our world view and materialism to the hegemony of the left hemisphere.

Apollonian Power and Dionysian Life
It will have become obvious that the left hemisphere does not have a pretty relationship with metaphysics or religion, unless it is a religion that is governed by laws and rules. McGilchrist compares the difference between the two hemispheres with the distinction made by Nietzsche in human culture: the left hemisphere is Apollonian, the right one Dionysian. One could say that the former is characterised by rules, the latter by life. Thoughts become rigid in the left hemisphere; to bring them to life again they have to be taken back to the right hemisphere.

The right hemisphere feels perfectly at ease with the realisation that two opposite truths may both be right. Example: all human beings are different but, from another point of view, they are also all alike. Or: the original purpose of the brain was to automate physical processes and capacities, but as human beings we are capable of transcending this, which creates the potential of free will. For the left brain this goes squarely against logic: 'A' cannot be equal to 'non-A'.

The left brain easily suffers from cognitive dissonance; it cannot tolerate two incompatible concepts. For this reason the left brain does not like dissidence, for there is but one truth. When we remind ourselves of what the left hemisphere is interested in – to use, manipulate, categorise, avoid surprise and therefore strive to control – it is not hard to suspect a strong will to power in all of this. Thus we can see that all kinds of sometimes beautiful ideals, when severed from human nature, and due to an aversion of other ideas, have led to terrible consequences. Persecutions of heretics may be inspired by religion, but also by politics and even science. History has examples galore in the Inquisition, crusades and purges, whether from the Church, the Taliban, Isis, fascists or communists. We all know examples of fundamentalist intolerance. In my own profession, medicine, I have colleagues who cannot accept that there are more ways to heal people than the regular western methods, and who therefore want to prohibit those. This illustrates the lack of imagination of the left hemisphere.

But an excess of imagination of the right brain also has its problems, such as a lack of critical thinking, conspiracy theories, all forms of swindle and fraud (also in medicine), fraudulent science, criminal gurus and suicide sects. In hippie days, the preeminent time of the right brain, the use of drugs increased tremendously. Throwing over all taboos and rules in those days at the same time wiped out many reasonable practices. Here again, we can all come up with our own examples. In the 1960's I saw the radicalisation of the student rebellion in both directions. Many students went to the extreme left, they followed their left brain and joined communist organisations such as the Rote Armee in Germany and the Red Brigades. The other hemisphere went via the hippie movement for 'expansion of the mind' through drugs. I have seen some brilliant minds totally lose their way by using LSD, and they never recovered.

Thus in their one-sidedness, the two hemispheres each lead to their own form of evil if we don't strike a balance between the two. And who has to strike the balance? I have to do that, *I myself*. Perhaps we could say that only then can I realise myself as my true self when I do not let my brain determine me, as happens in a compulsive disorder or addiction.

The Influence of Culture

The dominance of the left hemisphere is therefore not naturally given. In small children all the options are still open. For instance, they have been shown to be able to distinguish all the different kinds of sounds of the languages spoken in the world.[10] A six months old Japanese child can still distinguish the *r* and the *l*. After a year it no longer can. This would mean that our perception of sound is culturally influenced. But it goes much farther than language. Even what we see and how we see it is culturally determined. Different cultures not only produce different views of the world, but also perceive the world differently. This means that the cognitive architecture of the brain, the circuits, are different,[11] or rather, have become different under the influence of culture.

A commonly held view is that westerners approach the world analytically and easterners holistically, the domains of respectively the left and the right brain. Social psychologist Richard E. Nisbett has confirmed this with his research. Both in Japan and in the USA, a student of his showed eight coloured cartoons of swimming fishes to test subjects (also students). Every cartoon had one 'focus fish' that moved more rapidly and was larger and more prominent than the surrounding fishes. When the test subjects were then asked to describe the scene, the Americans invariably began with the 'focus fish.' Seventy percent more often than the Americans, the Japanese described the surrounding fishes, the rocks in the background, the plants and other animals.

Subsequently, they were shown separate fishes or other objects and were asked whether they recognised them from the films. The Americans thought they recognised them whether or not they had actually figured in the films. But the Japanese were much better at this; they remembered the surroundings of such objects. Then they were asked to answer as quickly as possible the same question when the objects were shown against a new, different background. Now the Japanese made mistakes and the Americans did not. The quick reactions tell us something about the automatic neuronal processing of perceptions (system 1), because the conscious mind (system 2) can then not yet have interfered. This proves that the perception circuits of Japanese are truly different from those of Americans.

And it suggests that Japanese rely more on their right brain, and the Americans on their left. Unfortunately, as far as I know, this has not been researched with the aid of imaging techniques.

Nisbett's team proved that this is not a genetic but a cultural difference by demonstrating that, after they had lived in the USA for a few years, Japanese were no longer different in their perceptions from Americans; and the same was true for Americans in Japan. And this was not due to their having been introduced to another way of thinking, but simply to being immersed in another culture.[12] Culture determines the way we think. Isn't it interesting that our world view not only influences our behaviour but even our perception? We are not so free after all.

Not only in our childhood do we use both hemispheres more equally, also when we age both hemispheres go into action with activities that are still strictly separated in the prime of life. Prefrontal activity that used to take place in just one of the two hemispheres occurs more and more in both as we grow older. See the HAROLD model of the preceding chapter.[13] *In the course of life, therefore we adapt the activity of the two hemispheres of our brain to the demands of the life we lead.* Ideally, the result of this situation is that one-sided intelligence makes room for the wisdom of old age.

The fact that the two hemispheres each give us their own view of the world is actually a wonderful thing. It enables us always to regard something from both (or all) sides.

East and West

Analysis means dividing a thing into its parts, and analytical thinking thus naturally leads to reductionism and materialism. Chinese physicists are more interested in fields and forces than in parts and particles. In China, magnetism and acoustical resonance were known long before they were in the West, and the influence of the moon on the tides had been discovered much before it was in the West. Also in medicine we can observe the same geographical difference in the tendencies to holism or reductionism. The French discoverer of the HIV virus,

13. CONFLICT BETWEEN NEIGHBOURS?

Luc Montagnier, who received a Nobel Prize in 2008, was a celebrated scientist until he discovered that homeopathic solutions can be effective. He now works in a laboratory in Shanghai, where his holistic views present no difficulties. It is fascinating to realise that for the way we observe and think it is not only important where we grow up, but also into what environment we move later on. This certainly relativises both views. But what does this mean? Are we completely at the mercy of the way of thinking in our culture or do we have anything to say about it ourselves? How do we arrive at judgments and decisions?

CHAPTER 14

The Role of the Brain in Thinking: Interim Summary

We are automatons as much as we are spirits.
Blaise Pascal (*Pensées*)

When we view Rodin's well-known sculpture 'The Thinker' our first impression might be that we are looking at a muscular sportsman rather than a thinker. And we do not usually attribute deep thinking to a sportsman. This is undoubtedly a prejudice, but the sculpture makes one thing very clear, namely that in order really to think, one cannot spend one's energy on anything else that requires brain activity – which, of course, is needed for everything we do. We have to concentrate all our energy on our thinking. Thinking takes more energy than other brain tasks. I once heard sprinter Dafne Schippers say: 'One thought, and already you lose speed.' Evidently, thinking is hard work, and as we know, the brain demands more energy than other organs. Now, what exactly is the relationship between thinking and the brain?

In previous chapters many facts have come to light about what the brain does and doesn't do. But facts become only interesting in their context. In this chapter I want to bring them together into a coherent picture of the role of the brain in thinking.

Expertise as Automatism

First of all, I hope to have made a credible case for the view that the brain is an instrument, namely the instrument of consciousness. And therefore also the instrument with which we think. The instrument may be of better or lesser quality, for it is our brain that determines our intelligence. This will not surprise many people.

Besides its innate quality, this intelligence is also dependent on what we have learned. As we have seen, thanks to the well-known plasticity of the brain, we have formed our brain with what we have learned. The interesting thing is that, after an injury, this same plasticity that enables us to learn also makes a potential recovery of the brain possible. Thus, plasticity has a threefold function; it enables the brain to function *and* grow *and* recover.

The manner in which the brain lets itself be formed by consciousness has been described down to the molecular level. The brain makes it possible for us not only to learn, but also to automate. In this regard, Victor Lamme is right: the brain functions like an automaton; one could say that it forms our automatic pilot. And this is not the same as saying that *we* are automatons. Owing to this characteristic of automation we build up expertise.

What is the key characteristic of expertise? It is the ability to quickly draw conclusions and make decisions in the field in which we are experts; or to make movements we have practised, not only in sports, ballet or acrobatics, also in music or trades. We are able to do this because we are familiar with patterns and can recognise them, also in movements.

To speak a sentence in our mother tongue we do not need to search for each word with much effort. The words come automatically. A chess master does not need to plan a lot of moves ahead of time, because he recognises the pattern of the positions and knows which move is the right one in that context.[1] Expertise is therefore in large part pattern recognition, and our association cortex is eminently suitable for that. Thus, not only do we see, hear and feel with many nuances, but most of all we recognise patterns.

Part of this recognition process is that we draw conclusions as to their significance. Without pattern recognition we would be lost in this world. The automatic pilot is the 'smart unconscious' of Ap Dijksterhuis' book.[2]

Fortunately we are not limited to learning from our own experiences, we can also learn from others by imitating them, not always visibly, because of our mirror neurons. The fact that we learn from others, whether we realise it or not, makes culture possible. And we not only learn from others, but we can also understand them by the same mirror neurons, which can evoke feelings in us that resemble those we see in them. It is the basis of empathy.

Jumping to Conclusions

This pattern recognition may also mislead us, however, when we think we recognise a pattern that is not appropriate or applicable. Think of visual illusions or personality disorders. The last is the result of strategies one uses in youth to avoid pain that is 'imprinted' in the brain so that we use them even when it is no longer functional. Dijksterhuis showed examples of times when the unconscious is not smart at all, but draws incorrect conclusions. As we have seen, the person who really explored this subject is Daniel Kahneman,[3] who called this form of jumping to conclusions *fast thinking* in contradistinction with conscious thinking which he called *slow thinking*. He also called them system 1 (fast, unconscious) and system 2 (slow, conscious) and pointed out that we prefer unconscious, fast 'thinking' because it takes no effort, while the slow kind does. Fast thinking is really not thinking at all, but automatically recognising patterns and making associations, which the brain 'does by itself'. The classic example is of course prejudice. This is almost a definition of system 1. Another example is Twitter: when you think too long about it, it is no longer current, with the result that many tweeters in retrospect regret their reactions.

We all suffer from prejudices. Even the most enlightened, liberal souls have been proven to harbour racial prejudices. This can be discovered with the aid of smart psychological techniques.[4] But that does not mean that they necessarily have to determine our words and actions, for we do not need to obey such prejudices if we switch to system 2. To really think, therefore, we must take a step away from what the brain dishes up as 'an offer you can't refuse.' We should use our brain only as an instrument. We don't much like to do that because it takes a fair amount of effort. Why does it take effort? And how is it at all possible to step back, and look at our brain as a separate object, as an instrument?

Ego Depletion

The person who researched why this takes effort is Roy Baumeister in his book *Willpower*.[5] He noticed in his research that people possess a limited amount of self-control – some more than others – and that, after having made many decisions, exercising self-control for an extended period of time, or performing many difficult tasks – which can all be viewed as forms of slow thinking – this amount of self-control becomes exhausted. He calls it ego depletion. Blood tests showed that this was accompanied by a lower blood sugar count and thus had indeed taken more energy.

Why would it take more energy? It seems as if it has to do with the fact that the person is not following the script of pre-formed circuits, but that new ones have to be formed. When we learn something new, which is only possible in the form of slow thinking, a much larger part of the brain is engaged than when what we have learned is often practised and becomes automatic. The latter requires a much lower number of neurons, which one could view as a sign of efficiency.[6]

Interestingly, this ego depletion is not found by all researchers.[7] But it has been confirmed, among others by Kahneman, that self-control and 'slow thinking' take energy. Therefore, Baumeister views slow thinking, which means that we don't let the brain think for us, but we do it ourselves – as if separating ourselves from it to be able to use it as an instrument – as well as exercise self-control (willpower) and make conscious decisions, as functions of the 'I.' *Evidently, it is the 'I' (or self) that can use the brain as an instrument.*

It is no coincidence that we speak of 'self-reflecting' consciousness. It is the 'I' that is able to determine whether the talents of both brain hemispheres are going to be used. It is also the 'I' that is the 'free willer' who is capable, to a certain degree, to determine his own behaviour, in spite of the well-intentioned suggestions made by the brain. It is the 'I' that enables us to prefer a new thought over an old one, because we consider the one good and not the other. It makes morality possible.

This would have to mean that the 'I' (or the *self*) is not created by the brain and, like consciousness, is not of a physical nature. It is something that does not apply to animals, for we do not hold animals responsible for their behaviour. They lack the potential of a will that differs from the brain. Animals follow their biology. As I have tried to demonstrate in Chapter 9, only the human brain offers this freedom so we can rise beyond its biology. At the least, this requires the extensive human prefrontal lobe that can silence the inborn and the learned automatic responses of the association cortex. Intelligence therefore does not suffice to guarantee responsible actions; these also demand a certain strength of the 'I.'

Therapists and Researchers

The last guest of a popular Dutch interview show on television in 2015 was psychiatrist Damiaan Denys. He said of his patients (primarily patients with obsessive-compulsive disorders): 'These are indeed people who *really are their brain*. Normally, we also have a spirit that stands above it and exercises control over things.' Actually, this is true for all psychiatric disorders, such as addictions and the various disorders that can end in psychosis. However, it does not mean that these people no longer have an 'I,' but their instrument doesn't obey this 'I' for some reason. Most people who were able to overcome a psychosis were demonstrably 'present' all right in that period, but were incapable of intervening. The brain had taken over, resulting in a painful crash of system 1 (fast thinking). All psychiatric and psychological therapy comes down to the objective of offering the 'I' or, in the words of Denys, the spirit, the opportunity to take charge of its own brain again.

In some cases, for instance, depression may drive a person to an inevitable suicide, because his or her thoughts of worthlessness can no longer be corrected by system 2. When Dutch author John Zwagerman ended his own life, psychiatrist and suicide specialist Jan Mokkenstorm was interviewed. The latter related that in his early twenties he had also walked around with intentions of suicide. To the question of the interviewer as to how he had overcome this, he replied: 'When I came to the insight that I was not my depression, but I *had* my depression.' This could be translated as 'that I was not my brain, but I *had* my brain.' This is the first thing a therapist would need to point out to a patient. Otherwise it would be impossible to enable someone to see perspective again. Therapists who think that we are our brain therefore have a problem. I do have to add that there are not many such therapists. We see this kind of thinking primarily in neuroscientific researchers, who do not practise therapy.

Supernatural?

If we are not a thing, in this case our brain, we must be something else. Who or what am I then? What is the 'I' or the 'self'? Has its existence become sufficiently credible? The 'I' cannot be scientifically demonstrated. Aren't we then opening the door to the supernatural? How can evolution have caused this? We might ask the same question regarding life. We still don't know what brings an organism to life. And consciousness? This also cannot be explained out of natural science. And the fact that we are able to reflect on our consciousness is just as much a mystery. That something which is not of a physical nature can exert influence on the physical world is not any more supernatural than the reverse: the idea that something physical would be able to produce something non-physical such as consciousness. And in practice this influence of consciousness is utilised daily in the form of games and prostheses that can be directed by the power of thinking.

Or, which is even much more important in this connection, the fact that counseling can cause demonstrable changes in the brain.[8]

But how can we make a credible case that consciousness (or the 'I') is not produced by the brain, for instance, in the prefrontal area, as many neuroscientists suggest? In situations in which the brain clearly is not able to function properly, while we can still witness the presence of intelligent and self-reflecting consciousness, we would have to presume that the brain is not causing this. The example of Bach-y-Rita certainly falls in this category, but there are even much stronger examples, which we will discuss in the next chapter. When the healthy brain directs consciousness we see system 1 of Kahneman in operation. When in a case of a seriously injured brain clear consciousness is still present, this cannot be due to the brain. This has to be something other than biology. This indicates system 2, or the intervention of the 'spirit,' in the words of Denys, a sign that the 'self' is involved here.

Part 2

I'm The One Who Thinks

'His brain forced him to take note only
of reductionist books about the brain.'

CHAPTER 15

To Be or Not To Be

Nature in her unfathomable designs had mixed us of clay and flame, of brain and mind, that the two things hang indubitably together and determine each other's being but how or why, no mortal may ever know.

William James[1]

One day British neuroscientist Adrian Owen was asked whether he could name a practical clinical application of brain examination by fMRI scan that is helpful to the patient. He could not think of any,[2] until he had the idea to use such scans on people of whom it was unclear whether they still possessed consciousness after damage to their brain. Maybe an fMRI scan could be helpful in discovering whether a person still experiences anything. There are actually four kinds of diminished consciousness possible after a brain injury, in most cases caused by a stroke, and sometimes it is difficult to determine which kind one is dealing with.

First, there are the locked-in patients. They are fully conscious, but due to damage to the brain stem they can no longer move their muscles, including those of mouth and face. They can therefore not communicate and, as a result, it is easy to think they are not conscious. They can only move their eyes and eyelids. This makes limited communication possible, for instance, by agreeing that lowering the eyelids means *no* and looking up means *yes*. Once it has been established that it is a case of a locked-in patient, consciousness is no longer in doubt. But the diagnosis is not always obvious.

Second, there are patients in a state of minimal consciousness. These patients have moments when they react to stimuli; they are in a dream-like condition that otherwise much resembles the vegetative state described next.

Third, the patients in a vegetative state. They appear to be awake, their eyes are open, they show reflex movements, but no reactions to questions or stimuli. Communication is impossible. There is no sign of any consciousness.

Fourth, the patients in a coma. These have their eyes closed and show no reactions whatsoever.

It is often very difficult to arrive at the right diagnosis in these cases and to know whether the patient still possesses any consciousness. In the first fMRI examinations by Owen on a patient in the vegetative state it looked as if the blood flow in the brain showed reactions to questions.[3] Critics did not consider this as proof; they said it could mean anything, such as a reflex reaction or plain coincidence.

Around the same time (2005–6) Belgian neuroscientist Steven Laureys had the same idea. The two scientists heard of each other's work and began a collaboration. They agreed to ask patients first to imagine they were playing tennis, and next that they were walking through the rooms of their house. When these questions are asked of healthy test subjects, the supplementary motor cortex (Figure E on the inside back cover) is activated in the first case, and in the second case the area near the hippocampus (Figure D) that is related to orientation in space. There proved to be only five out of fifty-four patients with serious brain damage who could react in the standard way. These cases too were judged by some critics to be showing an (unconscious) reflex. When healthy people react automatically to a question, this belongs to system 1, which unquestionably is part of ordinary consciousness.

Then Owen and Laureys had the bright idea to agree with their patients that tennis means *yes* and walking in the room *no*.

They asked 'closed' questions that could be answered simply by yes or no. One of these patients was able to give correct 'answers' to five of the six questions asked, even though she showed no visible sign of consciousness. For example, when asked about the name of her father she could answer yes at the right name. But it clearly took the patient a lot of energy. She could not concentrate for more than 30 seconds and then needed 30 seconds rest.[4] That does not look like a reflex, because a reflex needs no concentration. It definitely seems like involvement of system 2: a strenuous effort to be boss in her own brain.

A BBC documentary about this experiment showed that the mere fact of the discovery that one patient in a vegetative condition was able to answer questions was the cause of more intensive communication with him; eventually the patient was even able to go home and to communicate via a tablet with the alphabet on his lap on which he indicated the letters of the words he wanted to say. To the question of how he felt before the first scan was done, he answered: 'As if I was shouting against a wall.'

Are we witnessing consciousness here? For the patients who give the right answers we have to conclude that they possess consciousness, for we are observing a form of communication. This form of brain failure apparently does not always preclude consciousness.

Terminal Lucidity
There have been observations concerning consciousness that absolutely do not fit into any neurological theory. For instance, there is a remarkable phenomenon of which I long thought it was a myth. I am referring to demented individuals, or people who are otherwise handicapped by brain disorders, who in the last hours, or even weeks, before their death suddenly recover their normal mental and cognitive capacities and memories. Articles about this[5] and a book[6] have appeared in which dozens of different cases were collected. They contain descriptions of schizophrenic patients who, even after having been in a catatonic state for decades, suddenly showed normal behaviour again in the last days before they died.

The same was described of demented patients. The patients were able to speak with their caretakers and visitors, and make preparations for their burial and the disposition of their estates. Some examples:

A violent former lieutenant of the Royal Navy who suffered from a psychic disorder had lost his memory; he could not even remember his own first name. The day before he died he suddenly became totally rational and asked for a priest. The patient spoke with him with full attention and uttered the hope that God would be merciful to him. An autopsy revealed that his skull cavity was so full of a straw-coloured liquid that the ventricles had become greatly expanded, while the little brain tissue present had hardened.

The second example: the person who reported this case had a brother with a serious psychiatric disorder in an asylum. One day he received a message from the director of the asylum in which he was told that his brother wanted to speak with him. He immediately went to him and was astounded to find him in a totally normal state. When he left, the director told him discreetly that his brother's sudden lucidity was a virtually certain sign that his death was near. This proved to be true. Here also an autopsy was performed, and it showed that the brain had changed completely into pus and that this condition must have existed for a long time.[7]

More recently, a case was described of a patient with meningitis who was 'seriously disoriented until just before her death … but cleared up only a few minutes before her death; she answered questions, smiled, was slightly euphoric and came to herself.'[8] In 1975 two psychiatrists reported three cases of schizophrenic patients who went into remission and could therefore communicate normally just before their deaths.[9]

Such terminal lucidity may also occur in cases of dementia. Three cases were reported in 2004. In every one of these, the patients had not recognised their relatives for years, but just before their deaths they became normal again and recognised them again.[10] A nursing home physician told me she had a patient with serious Alzheimer's and metastasised cancer who went into a terminal coma.

After a few days she suddenly woke up and asked her children who were present to ask her former husband to come and see her. The children were surprised by the uncommon autonomy and energetic way in which she was organising her parting. When the man was there she thanked him for all the good things they had had together. An hour later she died.

These are merely case descriptions, but we can discover a certain consistency in the stories when we compare them with the research I will discuss hereafter. I have myself been present with a dying person who, because of widespread metastasis in the brain, had been unconscious for a long time, and became totally lucid for a short time before her death. And when I tell such stories to a hall filled with caretakers, there are invariably a number of nodding heads, and some listeners will report that they have witnessed similar events. It is of course a difficult subject to research with large groups of patients. We will always just see a few individual cases. But all these cases put in doubt what we thought to be so absolute, namely the interdependence of brain and consciousness.

Do we really need a brain?

In 1980 an article appeared in *Science* with the provocative title *Is Your Brain Really Necessary?* It contained a report by John Lorber, professor of pediatric and infant neurology, on his fifteen-year research into hydrocephaly (an abnormal accumulation of cerebrospinal fluid) in children and adults.[11] The article begins with a young mathematics student with an IQ of 126 and a verbal IQ of 134, who had the highest grades in his class and was socially completely normal. Because the university physician considered his head a little larger than usual, he referred him to Dr Lorber who performed a brain scan. He found that in lieu of the normal 4.5 centimetre brain tissue that should exist between the wall of the ventricles and the surface of the cortex, there was only a thin membrane of about a millimetre. It was difficult to ascertain the weight of his brain tissue, 50 or 150 grams, 'but it is evident that it does not even come close to the normal

1.5 kilograms.' His brain was filled primarily with cerebrospinal fluid.

He was one of more than six hundred cases of hydrocephaly collected by Lorber. Lorber developed this interest out of his specialty in *spina bifida* (cleft spine), a condition that often goes together with hydrocephalus. Cerebrospinal fluid is continually generated in the chorioid plexuses in the ventricles of the brain and circulates through the ventricles all the way down the spinal cord and back up again along the outside of the spinal cord to the outside of the brain. It is then absorbed into the membrane called the *arachnoidea* and relinquished to the blood. When somewhere on the way there is a blockage, the fluid cannot be drained off and the ventricles dilate due to the increasing pressure of the fluid. When this occurs early in youth, the skull also becomes larger. This happened more commonly in the past. Now these cases occur much less often because of better checkups during the pregnancy and thereafter.

Sometimes, however, hydrocephalus is found in a skull of virtually normal size, an 'internal hydrocephalus.' In most cases this leads to obvious handicaps, but sometimes it doesn't. Lorber divided his cases into categories of increasing 'water volume.' Of those who have 95% cerebrospinal fluid in the skull cavity – comprising 10% of the six hundred cases – one half are badly handicapped, but the other half have IQ's of over 100! These latter cases probably differ from the former, the handicapped ones, in the fact that they developed much more slowly.

Another group Lorber studied are those with one-sidedly enlarged ventricles. He has seen more than fifty of such cases and there were only a few that showed the expected spastic paralysis on the opposite side. The case of one of the one-sided paralytics defied the most certain knowledge we believe to have of the nervous system, namely the crossing of the nerves of our body. The paralysis was located on the same side as the enlarged ventricles.

Lorber owes his knowledge of these cases exclusively to brain scans. Because the persons were still alive, nothing was known about the structure, macroscopic and microscopic, of the brain tissue. In cats in which hydrocephalus was artificially generated, the loss of tissue occurred primarily in the white substance and almost none in the grey substance. This means that the number of neurons has not diminished, but the 'isolation' (by myelin) of the axons. Myelin is always considered indispensable for quickly conducting stimuli, and therefore for intelligence.

Since the appearance of Lorber's article, more cases have been published, such as that of a French government employee, married, two children, who felt a certain weakness in his left leg. His IQ was only 75. A CT scan and an MRI scan showed the following picture:

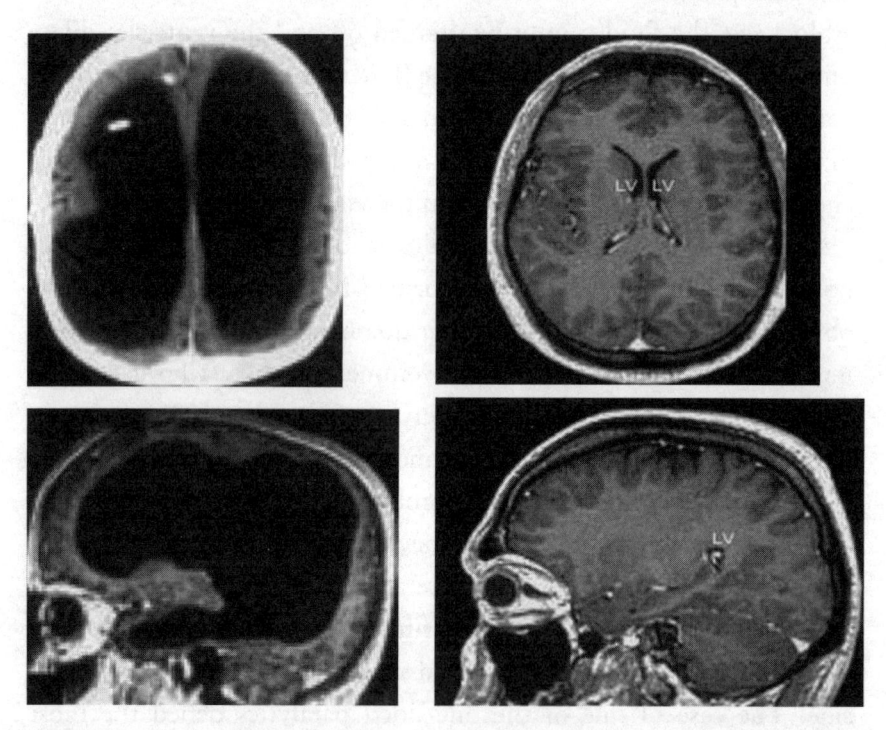

Figure 13. Left: the pictures of the patient. Right: normal pictures for comparison

There was also a case of a truck driver who had always been healthy, but when he crashed his truck against a tree, he went into a coma. A brain scan revealed that on both sides the entire frontal lobe and a large part of the parietal and temporal lobes were absent and had been replaced by a large cavity filled with fluid which, as shown by the picture in the article, occupied the whole front half of the skull. He was probably born with this, and no one had ever noticed anything wrong with him. He was put on oxygen and recovered completely.[12]

These kinds of observations are published quite regularly. So far, there have been no explanations.

One half suffices

There are additional examples of massive absences of brain tissue which, however, seem to have no notable consequences. In the course of time, neurosurgeons have tried a variety of surgical interventions in cases of untreatable epilepsy. Callotomy, cutting the corpus callosum, which results in split-brain patients, is no longer performed. But what is still sporadically done is hemispherectomy, the removal of that half of the brain in which the source of the epileptic seizures has been localised. It is performed on children under ten years and, remarkably, this does not seem to reduce their cognitive capacities.[13] If they are young enough even the removal of their left brain does not seem to result in speech problems. Only they no longer see the right half of their field of vision, and the right side of their body may be paralyzed.[14]

The following story, which does not come from the scientific press but from the internet, and is therefore potentially apocryphal, agrees with the sources quoted above. A little boy of eight with Sturge-Weber syndrome, which often goes together with blood vessel disorders in the brain resulting in epilepsy, was badly retarded in his development; before his surgery the only word he could say was Mama. After the removal of the affected brain hemisphere he experienced accelerated development and was soon fully normal.[15]

The left half of the brain of a three year old girl (daughter of Turkish immigrants) was removed in Holland. When she was seven, she spoke fluent Dutch and Turkish and showed only slight spasticity on her right side.[16]

Dying

What do all these cases have in common? It appears that the role of the brain, of which I have tried to demonstrate in this book that it is secondary to consciousness, and has the task of facilitating consciousness, is perhaps more enigmatic than we thought. Is there actually an indissoluble relationship between the biological function of the brain and consciousness? In regard to the event of dying, terminal lucidity is not the only phenomenon to expose the enigmatic nature of this relationship.

General physicians in Holland often stand at a person's deathbed and sometimes are present at a death itself. What particularly struck me time and again was the riddle of the relationship between biology and the moment of death. Even when we have an impression of the state of the body of a terminally ill person, still it is very difficult to predict when the person will die. There seems to be no absolute connection between the dying of the organs and tissue and the moment of passing. One person dies when the body does not yet give an impression of being unable to harbour life any longer, another dies at a time when we have the feeling that the body would have given up long before.

There are well-known cases when an ill person holds death back until a particular key person has come to his or her bed. And for others the event of death takes place when they have accepted that the time has come, even though it seems that they could well have waited a little. It is as if the biological process of the death of the body does not inevitably determine the moment of the departure of consciousness, or rather of the *self*. Sometimes such a person 'comes back' before departing, as shown by the reported cases of terminal lucidity (see above).

What is dying actually? Isn't it simply the consequence of the body giving up? Progress in medical science has called attention to two notable circumstances of dying that compel us to formulate our concepts of dying or passing away more precisely.

Brain Death

The first condition I want to discuss is brain death. The situation is here as follows. The body is kept alive – for otherwise no living organs can be taken out, but the brain is deemed to be so badly damaged that recovery is out of the question. The presumption is then that the owner has 'departed' and cannot return because the brain makes it impossible.

Another assumption is that the brain is indispensable for the regulation and maintenance of the integrity of the organs in the body. But this has been shown to be untrue. Alan Shewmon, pediatric neurology professor, has done research into the occurrence of what he rather ironically calls 'chronic brain death.' He published an article on it in the journal *Neurology*.[17] He began this study because of two cases he witnessed himself. One involved a fourteen-year old boy with a skull trauma. He remained in intensive care for several weeks and was then taken home where he received oxygen and vasopressin (antidiuretic hormone) and parenteral nutrition. He died after 65 days of an untreated infection.

The other case was an 18-year old boy with haemophilus influenza meningitis with such a high pressure in the skull that the seams of the skull split open. At the time of the article – fourteen and a half years later! – the boy was still alive; even with a flat EEG, without spontaneously breathing or brain cortex reflexes. Angiography showed no blood flow in the skull, and in imaging pictures the entire skull cavity proved to be filled with fluid and membranes. He was fed at home by means of a gastrostomy (in the stomach) and received oxygen. The term 'chronic brain death' seems to be particularly applicable to this case.

The criterion used by Shewmon for chronic brain death was survival by a week or more after the diagnosis. He chose this standard because articles about brain death usually state that heart failure always occurs within a week of the diagnosis, based on the view that the brain is supposed to regulate the integrity of the life functions. Among 12,200 international sources, he found 175 cases of chronic brain death.

His conclusion was that the statement that brain death inevitably leads to cardiac arrest within 48 to 72 hours (longer in case of children) cannot be maintained. Nor can the statement that we need a functioning brain to maintain bodily integrity.

This kind of survival is actually not so very exceptional. We know that it is possible to keep a brain dead pregnant woman biologically alive until the baby is viable about which an entire textbook has been written.[18] To be exact, brain death is not really a diagnosis but a prognosis. Because what happens to a brain dead patient from whom no organs will be harvested? All treatment is stopped ... so the person can die. Shewmon does not mention it, but it certainly looks as if all his cases ended in a vegetative state.

Still, there are rare cases of people who wake up, even with undamaged cognition, in spite of precise observance of the brain death protocol. I know two such cases in Holland, one of which did not involve a diagnosis but only a discussion of the option of organ donation.[19] Worldwide it has happened a few times, even on the operation table, just before organs were removed.[20] Despite the various brain death protocols in the world that describe all the tests that have to be performed, it is evidently still difficult to be 100% certain that the patient will never regain consciousness.

But what is the story here? Is it a bad diagnosis? Maybe not. It is possible that the diagnosis was made in good faith and in accordance with scientific standards. But how can someone then come back into the land of the living? This can probably be explained by the latest developments in reanimation.

Sam Parnia, assistant professor in critical care and reanimation, leads a group doing research into resuscitation (reanimation). We now have so many techniques, he says, that people with cardiac arrest after a traffic accident or other violent force, or due to a drug overdose, can be brought back to life after more than six hours, and in time perhaps even 24 hours, after their death. Of course this does not apply to cases where a skull trauma has caused irreparable brain damage, or if cancer or infection has led to a physical condition that can no longer support life. According to Parnia, the rule that five minutes after blood circulation stops, and therefore oxygen shortage occurs, the brain definitively perishes, is not true. The brain may no longer function, but it is not dead. Dick Swaab's research group described in 2002 how in a cell culture brain cells of a deceased person, which had been removed eight hours after the moment of death, were able to remain alive for several weeks.[21]

The breaking down of brain cells can be avoided by much longer (mechanical) heart massage, more frequent defibrillation, and longer artificial respiration than has been usual until now. Hypothermia, the lowering of body temperature, can lengthen this time spectacularly, so that after much more than the usual five minutes people can wake up again without brain damage.

The most dangerous moment is the point when the brain receives enough oxygen again at the end of reanimation. Oxygen is actually toxic, unless the cells can take appropriate countermeasures, but after such a long time without oxygen they are not able to do that. In order to stay alive, brain cells go into a kind of hibernation, which makes them vulnerable to oxygen. The oxygen flow has to be very carefully restored, otherwise swelling and increased pressure will still occur in the brain – resulting in death or, at the very least, brain damage. The problem with brain death, says Parnia, is that we measure the function of the brain cortex, not whether the cells are still alive.[22] He wrote a book about it, *Erasing Death*.[23] His work with these people who, as he says, had *really* passed away in the usual sense of the word, not just *nearly* (outside the emergency room they would have been considered dead), has given him extensive experience with near-death experiences, which he takes fully seriously as reports of how it is to be really dead.

Parnia discusses two cases of apparently injured brains which created rather a sensation. The first is that of a firefighter who was working in a house of which the roof suddenly caved in on him. He was rescued, but upon arrival in the hospital he was already in a coma. He had breathed in smoke for too long so that his brain had been without oxygen for a long time. Cooling was not yet used at the time. Eventually he settled into a condition of minimal consciousness which lasted ten years. And then he was suddenly lucid again. He recognised his family by their voices (he had become blind) and asked: 'How long have I been away?' He wept when he heard how long it had been. Evidently he had not had any conscious experiences during that time. Soon after this he died of pneumonia. His sudden lucidity was probably due to a Parkinson medication he had received shortly before. It stimulates the generation of dopamine.

Neurologist Oliver Sacks describes the same thing in his book *Awakenings*[24] of patients who had sunk into a vegetative state due to Spanish flu and who woke up after having been given L-Dopa that has a dopamine-like effect.

The other case is of a man who had fallen into a state of minimal consciousness after near-drowning. He was eventually cared for at home by his wife. Every night he kept her awake by his loud groaning. At her wits' end she gave him one night the sleeping pill zolpidem through his stomach tube. He indeed quietened down and … woke up and could speak with her! After a couple of hours he faded again, but he came back whenever he was given a new dose of zolpidem.

The few people who woke up after a diagnosis of brain death had been kept physically alive for the purpose of harvesting their organs. It had the effect of reanimation on them. Brain death is after all not the same as biological death. A number of physicians and ethicists are of the opinion, in my view justified, that in such cases it is not correct to speak of death, but of organ removal from dying persons who had given permission to do that.[25] A highly critical 'white paper' of the American President's Council on Bioethics proposed the term 'brain failure' for these cases.[26] I can imagine that this causes legal problems. For then death can only be established as the consequence of the organ removal.

Near-Death

Via Sam Parnia we now find ourselves facing another phenomenon that, at the very least, puts question marks on the role of the brain in relation to consciousness. It is the most remarkable, and by now oft described, phenomenon of near-death experience. The fact that this occurs more frequently in our time is due our ability to keep the life processes going in persons who, without medical intervention, would have been considered dead. As a result, they are able to return to the land of the living. In these cases, in spite of badly handicapped brains – perhaps with a flat EEG or during a brain operation without blood flowing into the brain and hypothermia – people report memories and experiences that are invariably described as much more vivid than dreams or even their normal day-consciousness. A wide variety of neurodeterministic explanations has been advanced on these experiences. For instance, the lower brain structures that are still functioning are supposed to be responsible for these 'hallucinations'. I find that hardly convincing, because it is then really strange that, when I wake up in the morning and have had no problems with my brain in the night, I still cannot remember much of my dreams.

Neuroscientist Steven Laureys who, as described above, specialised in coma has done research into the question of the memories of coma patients and patients with near-death experiences (NDE's). He had established two control groups; one related their memories of important events such as births and marriages, while the other told of events they had dreamed.[27] The NDE patients remembered the most detail and emotions connected with them, and reported that their experience felt more real than 'regular reality.' The memories of 'regular reality' of the second group were also more detailed and emotional than the dreamed events of the third group.

For that matter, Laureys does not think of NDE's as spiritual events: 'I'm a scientist.' But 'it is this dysfunctional brain that produces these phenomena,' he said in an interview.[28] Again a dysfunctional brain that strangely enough, just as with terminal lucidity, makes lucid consciousness possible. But perhaps we should say: does not make it impossible.

Dutch cardiologist Van Lommel did prospective research into near-death experiences (this is a study in which no stories of NDE's were collected a long time after the fact, but everyone who had experienced a state of near-death or clinical death was questioned immediately after it had happened) and published an article on it in *The Lancet*.[29] He suggested that consciousness remains present even in cases when the brain no longer functions. In his subsequent book he dealt with every argument advanced until then that says that an NDE is the result of events in the brain, and he demonstrates that none of these arguments fully explain the phenomena.[30]

Neurodeterminists, who only admit consciousness as a product of the brain, were not at all happy with this. Sam Parnia also disproves the various theories that boil down to a shortage of oxygen or an excess of carbon dioxide, or other chemical processes that supposedly explain NDE's. All these conditions also occur without being accompanied by an NDE. And by the way, says Parnia, our everyday experience with normal reality also triggers chemical events in our brain, and this therefore indicates nothing about the reality content of NDE's. The objection that people who had an NDE evidently were not dead, for otherwise they would not have been able to tell about it, is at first sight of course entirely true. The bodily cells remained alive, else they could not have been reanimated. But their owner was indeed gone, passed on, exactly in the same way as occurs when a person dies. For death is also a process in which the bodily cells do not immediately stop functioning. In view of the experiences of Parnia, a reanimation physician, we should, to say the least, admit the idea that NDE's actually tell us something of what it is like to die. This is also the view of most people who went through an NDE; they no longer have any fear of death.

Thus here also we are witnessing a living body of which the owner seems to have disappeared for a short time, for instance, because the heart no longer beats or the EEG is flat, so that the brain apparently doesn't function. Bystanders therefore think the person has died, but the owner comes back. Of this latter situation, a no-longer functioning brain cortex, there are a number of examples. One is a woman, Pam Reynolds, who had an operation on a blood vessel in her brain. The brain was drained of blood because any bleeding would have been fatal. This was only possible if she was operated in a hypothermic condition. Of course, her brain cortex did not function in this bloodless state. She experienced an NDE. She saw the operation and the instruments used, she heard conversations of the operating team and met deceased loved ones in a world of light.[31]

In the meantime there are many NDE reports on the market. They all differ. Various elements, such as watching one's own 'soulless' body from above, the life panorama, the tunnel and the light do not figure in each report. But still, all subjects give the impression that something extraordinary is happening, something that yields up an essence we do not know in a state of health, wholly different from dreams or hallucinations.

The other case is that of neurosurgeon Eben Alexander, associate professor at Harvard with an extensive and successful scientific career, who had always viewed NDE's as hallucinations. Alexander fell into a coma due to bacterial meningitis. No life panorama, no out-of-body experience, just a coma lasting seven days during which his brain cortex was totally inactive. It was 'all pus.' The prognosis was: 80% chance of death, otherwise at best a vegetative state. But after a week he opened his eyes. He had had an NDE and said that with inactive neurons in his neocortex, his extremely lucid experience was not in conformity with what we think we know as scientific truth. He developed an idea of the brain as an obstacle for a certain form of consciousness, namely of 'true' reality. French philosopher Henri Bergson had the same idea. When we consider that most *idiots savants* owe their exceptional but isolated cognitive achievements to a brain defect that causes a low IQ so that they lag helplessly behind in every other respect, we might think that the brain inhibits at least as much as it makes possible. Alexander writes:

> But in my case, the neocortex was out of the picture. I was encountering the reality of a world of consciousness that existed *completely free of the limitations of my physical brain*. (Italics by Alexander) [32]

The most important reason for including the story in this chapter is the fact that during the NDE Alexander did not know who he himself was and had no memories of his life. Yet he experienced himself as an individual, although fully connected with the rest of reality as he experienced it. This abstraction and the absence of personal memories are things I have not encountered in other NDE stories.

The following may now have become clear: both in the case of NDE and brain death we are witnessing the departure, in the experience of the bystanders, of the consciousness of a person whose body is biologically still alive. Dying and 'departing' are evidently not the same thing. After all, a plant also dies, but we don't call it departing. The death of the body is a biological process; 'departure' is the disappearance of consciousness and/or what we could call the *self* or, the *person*. This occurs already before the cells of the body have died. Evidently, the connection may be broken. And subsequently, by exception and only when the body did not biologically die, the connection may be restored.

We may wonder whether we know enough about what is actually happening when a person dies. Is it simply that the heart stops, there is no more oxygen so that the brain can no longer make consciousness? Is this the end of everything? When we have become convinced that it is not the brain that makes consciousness, this explanation, in view of the above-related observations, is no longer absolute truth. We then face the question that cannot be asked within a materialistic purview, but is also asked by Parnia: Where did the one who reappeared so unexpectedly sojourn in the meantime? In these cases the brain has been unable to offer 'the person' continuity, and yet this same 'person' reappears. This question is valid both for terminal lucidity and for the two cases of minimal consciousness described above, as well as for the man who came back after receiving L-dopa and zolpidem and of course, for NDE's.

But even if we were convinced that the brain produces consciousness and the 'self,' we are left with a question caused by, on the one hand, the rather frequent return of people with NDE's and, on the other, the much rarer waking up of a brain-dead person. Evidently, the body a person has left behind, or that no longer generates a person, retains its integrity and identity for quite some time due to medical intervention. Otherwise the person would not be able to come back. If not the brain, what is it that has preserved this integrity and identity?

One thing is clear in these stories: the relationship between consciousness and body was really broken. All pain was gone. The word 'departure' to indicate death is therefore probably indeed quite applicable regarding consciousness.

Thus we are led to a characteristic of consciousness and 'self' that we have not been able to get into clear focus until now. We may have come to the point of concluding that consciousness and 'self' cannot be described in materialistic terms; however, these examples strongly suggest that the idea of consciousness and 'self' as products of the brain is no longer tenable.

When Does It Begin?

The complexity of the brain is always advanced as the argument that it is conceivable that the brain produces consciousness. If that were the case, it would be surprising, to say the least, that we can already sense something like consciousness in a foetus or embryo of which the brain is as yet too undeveloped to bring this about. Ultrasound has given us many new insights into the behaviour of embryos and foetuses. At this point it is generally accepted that babies can already learn things in the womb. They recognise the mother's voice and the melody of her speech.[33]

And after birth they have been shown to recognise children's songs that were sung to them during the pregnancy.

At which point does all this become possible? We know that the brain of a foetus is sufficiently developed to direct bodily functions around the twenty-eighth to thirtieth week. Around the same time the eyes can be opened. For convenience, let us assume that the brain is then sufficiently developed to make consciousness possible. But much earlier, from about the eighth week, when we are still calling it an embryo, movement can already be observed.[34] And around the eleventh week the foetus can make a fist. From the fifteenth to eighteenth week the baby already makes sucking motions, and from the nineteenth to twenty-first week it can hear and swallow, and the mother can feel movement! All these movements can still be viewed as reflexes that have no need for input from the brain.

Around the twenty-third week the baby starts to show REM sleep, probably including dreams, which would indicate a form of consciousness. Only around the thirty-second week quiet sleep begins. But there are also observations in the thirtieth week that presume 'real' consciousness. Around the twenty-sixth week the baby can already make movements indicating fright due to loud sounds.

Amniocentesis usually takes place after about fourteen to sixteen weeks. It has been shown that babies immediately react to this with much more movement.[35] There have even been observations showing that the baby repeatedly tried to push the needle away. This means that we are not witnessing spontaneous reflexes, but purposeful actions, notably in a phase when the eyes are still closed and the (motor) brain cortex has not yet reached its complex structure.[36] Since the brain cannot be held responsible for this, it has to be a form of independence of the brain. This might indicate the presence of a 'self.' But it is of course difficult to speak of an 'I' if there is no 'other.' The mother could be the other; there are many mothers who feel a relationship to their unborn child.

An objectively present other exists however in the case of twins. In such cases it already becomes clear in an early stage that the two babies may be quite different. Dr Birgit Arabin, perinatologist, has followed twins with the aid of ultrasound from the eighth week. They have been shown to react to each other already after nine and a half weeks.[37] Child psychiatrist Alessandra Piontelli has accompanied many twins from pregnancy onwards. She followed them with ultrasound from the eighth week. She describes how different twins are already from the first images in the womb, and how dissimilar their reactions to each other can be. This behaviour turns out to be indicative of their behaviour after birth.[38]

Sir William Liley, specialist in intra-uterine intervention, had already observed the individuality of foetuses ten years earlier and wrote an article in 1972 (*The Foetus as Personality*) in which he demonstrates that the foetus is an intelligent, acting personality that learns from, and reacts to, its environment, especially the mother of course.[39] We do not come into the world as a blank slate, but as a somewhat formed and developed personality, our own 'self.' It has been shown that the development of the brain in the womb is also preceded and formed by behaviour and consciousness, one could say: by a 'self.'

But we cannot act as if we have nothing to do with the fact that many people have the opinion that philosophy as well as neuroscience have done away with the idea of 'self' and 'I.' That is the subject of the following chapter. Consider it as a concession to the well-known objection of the left brain: 'So, it works in practice? Fine, but does it also work in theory?'

CHAPTER 16

Do 'I' Exist?

I am most certain that I am, and that I know that I am and delight in this.

Aurelius Augustinus

If a man will begin with certainties,
He shall end in doubts.
But if he will be content to begin with doubts,
He shall end in certainties.

Francis Bacon

James Fallon is a famous brain researcher specialising in brain scans and the genetic profile of serial murderers. He has become such an expert in this that after a quick look at a series of PET scans he can right away pick out the serial killers, for he says that they have reduced volume in the orbitofrontal (situated above the eye sockets) and temporal cortex.

His family shows a tendency to Alzheimer's, and for him, professional brain scientist, that was reason to make PET scans of the entire family. Fallon evaluated the results himself. To his horror one of the scans showed all the characteristics of a serial killer – someone in the family was a psychopath! But who? You know what's coming: that scan turned out to be of himself, James Fallon, compassionate Catholic, respectably married since his twentieth year, with children, good position, home in a nice neighborhood, generous *bon vivant*, had never hurt a fly. Moreover, his genetic profile looked suspiciously like that of a psychopath. His mother told him that there were murderers among his father's forebears. He asked his friends and family what they really thought of him. Did he look like a psychopath? Everyone thought he was a charming guy, but to a certain extent – he did not allow true, deeper contact. 'And you know what?' he confided to an interviewer, 'I couldn't care less.'[1]

Here he is. He has had so many interviews with photos, and he figures openly on YouTube, that I do not think I am invading his privacy in this way.

Figure 14. James Fallon

What is your impression of him? Charming and *bon vivant* without a doubt. Or should we not draw conclusions from someone's appearance? Wittgenstein called the body the best image of the soul, and I think that he is quite right. A person's bearing and manner of moving tell us a lot about the person's inner condition, but the face gives us perhaps more information than anything else. That is precisely what a brain scan can't do. I'd be surprised if a brain scan could give us such an immediate idea of charm or joy of living.

In brief, even if one is born with the cerebral equipment of a psychopath, like a genetic birth present, that does not mean that one is so determined by it that one indeed becomes one.

It is evidently possible to possess a sense of responsibility, even when the brain is not equipped for it, and to mobilise a considerable measure of self control. *This self control would have to come from the orbitofrontal cortex. But the reduced volume of that cortex was precisely the characteristic of psychopathy that Fallon discovered in his own scan.* It is interesting that Libet also found in his experiments that self-control, in the form of what he called 'veto power' – deciding not to do what one had decided to do – surprisingly was not preceded by any special brain activity. And according to Libet, self-control is the essence of free will ('free won't'). This could create the impression that therefore the brain is not the initiator of this.

But who then exercises the veto power? Isn't it I myself? Or in this case, Fallon himself. But what does this self mean? Where does Fallon's self control reside? In his orbitofrontal brain cortex that is too small? Where resides his *self*? Apparently, intact frontal lobes are not an absolute prerequisite for leading a normal life, witness the truck driver in the previous chapter, who even lacked a large part of his parietal and temporal lobes without anyone ever noticing anything the matter with him. He was born that way, but if his brain had been thus damaged later on, it would have been a different story. But psychopaths are also born with a defect in the frontal and temporal brain...

Does the Self Exist?

We are living in the twenty-first century, and we cannot be so naïve as to think that the *self* or the 'I' are not disputed concepts. What then is the problem with the self? First we have to come to an agreement as to what we mean with the word. A distinction is made between the *self as experience* – 'I am I,' a sweeping discovery made by virtually every child around the age of three – and the 'I' as *acting self* (an agent).

The self as experience may be subdivided into two kinds: the experience of one's own body – of the fact that one is embodied – and the experience of one's own consciousness. Together this is the experience of one's own person. Maybe one's name is also part of this.

However, there are reasons to have doubts as to the solidity of the experience. The *experience of the body*, for instance, turns out not to be reliable. An experiment that has to prove that even our consciousness of our body is a creation is found in the 'rubber hand illusion.' A rubber hand, which is lying on a table in view of the test subject, is caressed in the same rhythm as the real hand that is hidden behind a partition.[2] After a while, the test subject experiences the rubber hand as his own.[3] In monkeys it was found that their sensory and motor cortex reacted to the touch of the rubber hand.[4] *This is another indication that consciousness gives form to the brain, for no connection exists between the brain cortex and the rubber hand.*

The *experience of consciousness* is equally problematic. Philosopher David Hume (1711–76) showed in his time that in the experience of our own consciousness the self is nowhere to be found:

> For my part, when I enter most intimately into what I call *myself*, I always stumble on some particular perception or other, of heat or cold, light or shade, love or hatred, pain or pleasure. I never can catch *myself* at any time without a perception, and never can observe anything but the perception.[5]

This observation by Hume, which anyone can confirm, has led philosophers to doubt the existence of an 'I.' When we go in search of our 'I,' all we find are ever varying and arbitrary experiences. Does the self therefore exist? Or, applied to myself, do I exist? Hume says I don't, and many others have followed him in this view. But Hume was certainly not the first to wonder whether he actually existed. And this question is inextricably linked with the question of whether we are capable of having a reliable picture of the world, of reality.

Brain scientists describe how different areas in the brain represent different parts of the body or perceptions of the outside world. The question that must be asked here is: to whom do they represent these? It is a question that cannot be answered within neuroscience.

For then people soon fall into the fatal trap of the 'homunculus,' the idea that there is a little man in the brain who sees what the brain comes up with in the form of representations. As we have seen in Chapter 6, the problem here is that this homunculus would also have to have such a little man in his brain. It is of course a straw man argument, for who could ever maintain that there has to be a little man in the brain? It is a mechanical concept.

But all the information that has to represent a specific moment in time – the 'now' – comes in along different tracks that all have different time spans. The differences may be as much as half a second. How does that all come together as one whole? This question is called the binding problem. There is no location in the brain where this takes place. That location would have to be the correlate of the 'I,' which a number of brain scientist have been looking for in vain. For many a neuroscientist, such as Victor Lamme, this means that the 'I' and the 'now' are illusions.[6]

Doubts

Wondering whether or not we exist, provided it is not an academic-philosophical exercise, will usually end in a visit to a psychiatrist. But who of us has never, if only for an instant, whether or not 'under the influence,' wondered whether the world around us is actually real? Descartes even became famous because he had doubts regarding both questions: are my ideas of the world right, and do I actually exist? Aren't I dreaming? Aren't my thoughts and pictures of the world illusions that an evil demon inspires into me? Descartes' solution was: let me begin by doubting everything. What does that leave me? This: I doubt, therefore I think, therefore I am. Even if my thoughts and pictures are illusions, I am still a *self*.

What a relief! For then our daily intuition that we exist is not an illusion. But anyone who thinks that Descartes was the first with this experiment in thinking should read *The City of God* by St Augustine, the fourth to fifth century Church father Aurelius Augustinus. We can find there a startlingly similar line of thought,

16. DO 'I' EXIST?

a breathtakingly ingenious text, which follows on the first quotation at the top of this chapter:

> In respect of these truths, I am not at all afraid of the arguments of the Academicians, who say, What if you are deceived? For if I am deceived, I am. For he who is not, cannot be deceived; and if I am deceived, by this same token I am. And since I am if I am deceived, how am I deceived in believing that I am? For it is certain that I am if I am deceived. Since, therefore, I, the person deceived, should be, even if I were deceived, certainly I am not deceived in this knowledge that I am. And, consequently, neither am I deceived in knowing that I know. For, as I know that I am, so I know this also, that I know. And when I love these two things, I add to them a certain third thing, namely, my love, which is of equal moment. For neither am I deceived in this, that I love, since in those things which I love I am not deceived.[7]

In view of his education by the Jesuits, Descartes may have read this text. By the way, he denied this in a letter in which he thanked his correspondent for the tip to check Augustine for this. Fortunately, he is not yet happy with his 'I am' because the next question presents itself: what am I then? In view of the manner in which he discovered his own existence, it is not surprising that he concluded: 'Through thinking I can enter into my own existence; therefore I am a thinking thing.' With 'thing' he does not mean his own body, and thus certainly not his brain, for in his thought experiment the brain already belongs to the unknowable world, the world of which he does not know whether an evil demon deludes him with it. Rather, he means the point in consciousness out of which he looks upon this 'world.' The skeptical question of 'are we able to truly know the world, or do we only have a self-made picture of it?' has haunted philosophers ever since.

Personal Perspective

Thus there are two problems, and those are also the problems that are the subjects of this book, namely the philosophical and scientific questions of 'can I perceive the world or is it my representation?' and 'do I exist, and if so, what am I?' Both problems have everything to do with each other, because perceiving the world is something we do out of a personal perspective.

However, ever since Hume's idea that we are no more than a bunch of experiences, this personal perspective is for many, including philosopher Jacques Derrida and psychiatrist Jacques Lacan, no more than a story. Or rather, a web of stories with which we construe ourselves – an illusion therefore. That idea was eagerly adopted by Daniel Dennett:

> Thus do we build up a defining story about ourselves, organised around a sort of basic blip of self-representation. The blip isn't a self, of course, it's a representation of a self (and the blip on the radar screen for Ellis Island isn't an island- it's a representation of an island).[8]

According to Dennett, therefore, we have no real *self*. Or does he mean that we don't really know our *selves*, but merely a representation of them? That is something on which I could agree with him. But no: he calls the self a Centre of Narrative Gravity, which comes into being because we have language. We profess reasons for our deeds and demand reasons from others. We talk to ourselves, and this creates the illusion that we have a self. This develops in our youth. Our 'I' is nothing but a story we tell ourselves, and therefore it does not exist. 'I' does not exist. Nowhere in the brain can we detect a *self* where everything comes together.

'We create a story about ourselves.' When we stop and really contemplate this sentence, it becomes clear that even the denial of the existence of a *self* is not possible without a first-person-perspective: 'we' and 'ourselves.' Who creates or denies the self here?

And to whom is that story told? Consequently, we can't get rid of this self; it is the self itself that tells the story.

Theory of Mind

But there are neuroscientists who manage to answer both questions – 'Does the 'I' exist?' and 'Can we really know the world or do we only have a self-made representation of it?' – in the negative in one fell swoop. Victor Lamme is one of them. They say that we only have the idea that we are an 'I' because we wrongly think that the others are all persons who have their own motives. We get this idea because we are put together in such a way that we always look for other people's motives. We call this preoccupation *Theory of Mind* (ToM), the theoretical manner in which we try to interpret other people's behaviour.

According to Lamme, we also apply this projection technique, this ToM, to ourselves. As a result we suppose, wrongly according to him, that we are in a conscious way responsible for what we do, and we have the illusion that we are an 'I' or possess a self.

Neurophilosopher Alva Noë justifiably gets angry at the idea that people would first study someone's behaviour in order to then conclude that this other is a conscious being with his own motives. He calls this a sceptical aloof-scientific view of human culture, as if we only have access to someone's behaviour while their inner world remains closed to us by definition. ' No sane person can take seriously the suggestion that our knowledge of other minds is merely hypothetical'.[9]

The question 'can we know reality?' has become a problem due to the idea that we create a representation, a picture of reality in our brain, that we form a hypothesis of it in lieu of the idea that we form part of reality. This is a direct consequence of materialism. Perhaps it can well be defended within materialism, but our experience tells us a different story.

Brain-at-Rest

Or does the brain tell stories after all? In the beginning of this century a neuroscientific discovery seemed to indicate the place where the 'I' – 'as an endless web of stories about ourselves' – is hiding in the brain. It is not really a place where everything comes together, but a place where 'the self creates its story,' a place that appears to confirm Dennett's theory.

Neurologist Marcus Raichle explored what happens in the brain when for a moment we have not given it a task, when the brain is 'at rest,' like during a vacant stare or when we are daydreaming. It turns out that when consciousness has no focus, brain activity does not stop but becomes concentrated in a few areas of the brain cortex. And that area uses as much energy as the areas that are at work at other moments. Writing in 2001 Raichle called this the *default mode network* (DMN).[10] It is the standard network that becomes active when the brain has nothing better to do. We can call it brain-at-rest. It is pre-eminently a domain of the effortless, automatic system 1 of Kahneman. The more active this network is, the more loose thoughts someone will have without any immediate cause (see Figure F on the inside back cover). The active areas of the default mode network show as orange/yellow. On the left is the inside of the left hemisphere, on the right are the parietal areas. The activity takes place first of all in a part of the medial prefrontal cortex (which we use, among other things, to make a picture for ourselves of what another person experiences); then also in the medial area of the temporal cortex, the parahippocampal gyrus (which lies next to the hippocampus and plays a role in our episodic memory that we can therefore formulate in the form of a story). Then in the rear part of the *gyrus cinguli* and the adjacent cortex called *precuneus* (which plays a role in self-reflection and episodic memory (the story of one's life therefore), spatial processing and awake consciousness, see Figure D). And finally in the rest of the parietal cortex, which is used to associate sense impressions into spatial orientation. The largest area is therefore to be found on the inside of both hemispheres, deep in the gap between the two, and also on the inside of the temporal lobe, as if 'inner world' has to be taken literally.[11]

What is the mental activity the DMN gives occasion to? Daydreaming, thoughts coming and going all by themselves, the wandering mind, the stream of consciousness. Most of the time these are memories or plans, judgments of ourselves or of others, things we still have to do and things we have already done, well or poorly – everything that comes to mind when our attention is not specifically directed.

The researchers who first discovered this, think that the task of this network lies in preserving our own history through which we make our consciousness into our own personal consciousness. The hippocampus is always activated when the brain-at-rest is active. This indicates that memory is involved.[12] It is our own inner world untouched by outside stimuli or tasks we set ourselves. That closely resembles the 'I-story' of Daniel Dennett. It is also the area we use to chew on memories and meetings. 'What did John mean with his remark? What does Mary think of me?' And that actually looks like the Theory of Mind that Lamme points to as the source of the 'I-illusion.'

When the brain receives an order, meaning that we direct our attention to something, the DMN falls silent. Our attention is then completely directed to something else, not to ourselves. We then 'lose ourselves' in a task. Here also one area suppresses another, and vice versa, except with depressed people, the content of whose thoughts might best be described as a 'day-mare' – thoughts of worthlessness fill their consciousness and stay with them through all they do. In fMRI tests it has been shown that their brain-at-rest does not stop.[13]

At this point it won't be a surprise to anyone that the DMN is another feature that is formed only in the course of our lives; it is not a gift at birth. We form it ourselves.

Meditation

It has been shown that depressed patients can learn to stop this brain-at-rest, namely by a form of attention training called mindfulness,[14] a form of meditation. This means of course that we can ourselves stop an activity of the brain. Perhaps unjustifiably, I assume that many readers have at some time made efforts at meditation. Then we will notice that our consciousness tends in the direction of the brain-at-rest which takes it toward its own preferred, often trivial thoughts (system 1). As Hume had already observed, consciousness always looks for content. But as soon as we notice it, are *present* again with focused attention, we can intervene in the wandering mind (system 2).

We can then have a clear experience of the difference between consciousness and self. For the one who notices that the mind is wandering is we, the self. That is a form of attention and self-control. It means therefore that the brain-at-rest does not represent the place of the self in the brain, and also that the self is not a story. On the contrary, the self is the authority that can stop the story of the brain-at-rest. For who notices the trivial thoughts? Who stops them? It means that we can know our consciousness, observe it and even direct it. Experienced meditators, such as Buddhist monks, have shown an ability to focus their attention much longer than usual.[15]

But likewise it means that we cannot observe our *selves*. For it is the *self* which knows, observes and directs. We cannot truly know our *self*, just as we cannot call it on the phone. Hume rightly observed that in his time when he wrote that, when he went in search of himself, he only ran into perceptions. Without knowing and observing our consciousness we are unable to direct it.

Rabbinical Common Sense

The denial of the 'I' is a counter-intuitive theory to be ascribed to the intelligence of the left brain. We might wonder whether the 'common sense' of the right brain might be of some help here. In his novel *Humboldt's Gift*, Saul Bellow describes a philosophy professor with the name Moses Cohen. During a lecture a student stands up and asks him: 'Professor, how can I know that I exist?' The professor replies, as one would expect from a person with such a rabbinical name, with a counter question: 'Who is asking this question?' – the extensive deliberations of St Augustine reduced to a one-liner![16]

It illustrates the approach that can help us find our way out of this muddle. *Who* has Hume's loose experiences? *Who* thinks that the rubber hand is his? If the brain merely represents reality, for *whom* does it do that? For *whom* is the picture of reality a picture? *Who* creates the story about himself? *Who* is reading this text?

When we consistently ask the question 'Who does it?' and 'For whom does it happen?' we will recognise an 'I' in every subject discussed in this book.

We Ourselves Form Our Brain

First of all we found that, almost always due to interaction with others, we form our brain and change it. A *self* can only exist in interaction with others, otherwise it is an empty concept. (Fallon himself thinks that his loving upbringing made all the difference, which undoubtedly gave him his opportunity.) When these interactions have been unfavorable, resulting in addictions or personality problems, something needs to be mobilised that can take the reins in hand. This is the *self*. For in the end, we have to do it ourselves. This holds true for all of us of course in regard to all the things we don't like in our own behaviour.

Mirror Neurons

Mirror neurons (see Chapter 4) enable us to have empathy for another person, but they do not produce this empathy automatically – we have to do that ourselves. Mirror neurons are regular neurons with a different kind of connections that we caused *ourselves*. Our identity includes how we think about life. The view of life that is incorporated in our brain has, as we have seen, an influence on the way we behave. Upbringing and events in our youth also play an important role in this. Samuel and Pearl Oliner, both Holocaust survivors, interviewed Polish people who had saved Jews during World War II.[17] They all said that they did it because in their upbringing it had been inculcated into them to do the good. And yet, what they did was a lot more than acting normally and decently as was expected. Possibly they were too modest to truly name the role of their own selves.

The Supposed Illusion

However, no one has been able to locate this *self* in the brain. That is the reason why neurodeterminists assert that a separate self does not exist and is therefore an illusion. It is the brain that 'takes on the tasks of the self' (Dennett) and is therefore fully identical with the *self*. But logically that is incompatible with the idea that we ourselves change and give form to our brain, which is why this received the deterministic formulation of *the brain changes itself*. But the story of the Bach-y-Rita family in Chapter 2 showed us that when the brain is left to itself, nothing happens in it. Neuroscientist Joseph Ledoux therefore formulates it in the subtitle of his book *Synaptic Self* as follows: 'How our brains become who we are.'[18]

Many Selves

There is a school of psychology that says: we do not have an 'I;' we have a lot of them. We constantly play roles. But who plays those roles? Right! Was that also true for the woman who was both blind and seeing (Chapter 4)? She did not play her alter egos; that happened unconsciously. But doesn't that mean that her brain caused it all? Is that true, did the brain do that? An alter only needed to be addressed by name and it appeared. There was evidently a director present who heard the name and had only one task, namely to put the pertinent personage on stage. Who was this director? Who else but the woman her*self*? Irrespective of whether she was conscious of it or not.

Psychoses

No matter how confused someone may be, often there still proves to be an 'I'-function present. People who went through a psychosis can often relate precisely what they experienced and how frightening it was. It would never have been frightening if there had not been a healthy 'I'-observer through it all. I reported the following experience in my book *How Matter Got Spirit*:

> A patient called me in a panic. He no longer saw a difference between himself and a vase with flowers that was standing on the table. I asked him: 'How can you be in a panic if you are a vase with flowers?' The cause of his hallucinatory experience was an increase in the pain medication he received for his cancer. Later he said: 'Well, that proves again that we are just a bag full of chemicals.' I tried again: 'If that is so, why didn't you just accept that you were a vase with flowers?' 'Because I didn't like that idea.' 'Doesn't that then indicate that you yourself had not changed and was able to observe your changed experience?' That was an idea he *did* like.[19]

Conscious or Not?

This leads us to an interesting question: is the *self* conscious or not? Or rather: does the self only work in our consciousness? Dijksterhuis rightly said that the fact that we do many things unconsciously tells us nothing about free will. For whose unconscious is it? The unconscious (Kahneman's system 1) is ours, and we ourselves have largely given it its form. That's why the *self* evidently knows the way in the unconscious. It has made its way by itself. We recognise that in the phenomenon of the word that is on the tip of our tongue and doesn't want to come out. The word is then still unconscious, but at some time we brought it in *ourselves*, and we know that we know it. After some time, perhaps hours, it finally comes out. Is that just because the brain computer has performed its search function?

It is actually a strange phenomenon that we know that we know something while at that moment we cannot produce it. *Who* is the knower? In a normal conversation most of the words come out of our mouth without any delay, while we have made no effort to find them. But they come from the same unconscious source. Sometimes we only hear our own thoughts after we have spoken them – we call it thinking aloud. How conscious are these things? They are at the same time totally conscious, yet they originate in the unconscious. But one thing we know: these words come from *ourselves*; we totally recognise them as our own words.

Perhaps the brain first has to show some defect in its functioning before we discover that we *are* not our brain but we *use* it. An anonymous article appeared in the journal *Nature* of 6 November 2013 by a young neuroscience professor.[20] He relates that he was diagnosed with Parkinson's disease. He writes it anonymously because he does not want to jeopardise his scientific career by going public with it. For many of his colleagues think that Parkinson's disease irrevocably leads to cognitive deterioration. He describes how sometimes he cannot lift his arm while there is nothing wrong with the arm:

> But I have to put effort, even focus, into getting it to move — frequently to such a degree that I have to pause whatever else my brain is doing (including talking or thinking). … The way my mind and body do battle forces me to reconsider the homunculus, a typically pejorative (among neuroscientists) caricature of a little man pulling levers inside our heads, reading the input and dispatching the output. Virtually all that we know about how the brain is organised belies this image, and yet there is a dualism to my daily experience.

It is ironic that a neuroscientist, who here intuitively describes his *self*, has to do that anonymously.

16. DO 'I' EXIST?

But Where Does the Self Come From?

This non-physical character is connected with what may be the biggest objection to the existence of a *self*, namely that we human beings simply form part of evolution. If animals have no *self*, why would human beings have one? Where then does it come from?

In Chapter 9 we have seen that besides growth in freedom there is a second tendency in evolution in the form of individualisation. To understand the place of the human being in evolution, primatologist Frans de Waal's books are most helpful. He shows that empathy, sense of justice, altruism and care for each other – in brief, moral qualities – did not arise in the human being as a result of reason and civilisation. Rather, they are to a greater or lesser extent present in all social mammals, especially in apes. And that is a good thing, he says, otherwise we would constantly have to go against our nature. We can observe the same thing in the area of free will, a phenomenon that can be traced through all of evolution, but for which humans have the most advanced instruments, even if our free will is not yet very robust, as is demonstrated time and again by social psychologists. Thus we can also see that together with free will the potential of a 'free willer,' an individual, a 'self' has been prepared in the course of evolution by the individualisation developing in its course. But that 'self' of ours is also still rather fragile.

Betting on Two Horses

We have seen that the brain not only determines intelligence, but it even presents us with two kinds of intelligence. While the right brain sees reality as a whole, the left brain can only understand reality if it is just as familiar as man-made things. It understands things as models of reality. While the right brain recognises organisms and autonomous beings, the left sees things and models. On the one hand this difference is an evolutionary fact; on the other hand, it is reinforced or weakened by ourselves, since we always prefer the one or the other. But in human culture one of the two may already have been chosen so that it becomes virtually impossible to break away from it.

In science, the generally accepted preponderance of the left brain, which can suppress the right half, has caused thinking in our culture to develop in the direction of reductionism and physicalism. If we want to be able to go against generally accepted cultural and scientific opinions there has to be something that can take on system 1 with its ready-made opinions of the brain. This is the *self*. We have to *will* this. And to do this, the *self* needs to make use of system 2, conscious, slow thinking.

Self-control

We have seen that the pointers in the direction of an absence of free will, except in laboratory situations stripped of all context, are not convincing. Moreover, a belief that free will does not exist results in negative consequences for social behaviour. People who do not believe in it have a greater tendency toward unsocial and automatic behaviour. It is evident that free will has everything to do with *self-control*, a word that itself clearly indicates that it has to happen through the *self* which, however, appears to be more present in some people than in others.

But this does not mean that it depends on the genes. As Moffit and Caspi found (Chapter 10), identical twins may differ in it right from the start. As we have also seen, self-control takes a great deal of energy, just like attention (system 2). That they actually do take physical energy is proven by the fact that eating some sugar can put an end to 'ego depletion' or make system 2 possible, as demonstrated by Baumeister and Kahneman. (The 'I' needs blood sugar for its presence, one can say.) Because it takes energy, it is much easier to forget about self-control and attention. Maybe that is why self-control is often called a virtue; we have to actively do something for it. It takes effort.

CHAPTER 17

Afterword:

Is there an Alternative to Materialism?

There speaks the learned gentleman!
What you can't touch is way beyond your ken,
What you can't grasp with hands wholly eludes you,
What you can't calculate you think cannot be true,
What you can't weigh you say it has no weight,
What you don't mint you won't let circulate.

Goethe[1]

There is no such thing as philosophy-free science; there is only science whose philosophical baggage is taken on board without examination.

Daniel Dennett[2]

If we want to rely on science to understand reality, materialism is a scant resource. Not only does it keep many aspects of reality outside our purview, materialism also saddles us with an ethical problem. For if we base ourselves on materialism, as neurodeterminists do, deeds have no significance. Everything is a natural phenomenon. Deeds do not distinguish themselves from events, for the simple reason that deeds have no authors. The latter are but illusions. And the illusory author also lacks the freedom to direct his behaviour. Even this freedom is an illusion, which means that responsibility is also an illusion. There is therefore no longer any reason to admire anyone for his deeds or to hold him or her responsible. Spinoza liked that idea, for then we also lack reasons to hate people.

The most poignant illustration of this is given by evolutionary biologist Richard Dawkins in a letter he once sent to a newspaper with the following tenor. It is just as senseless to punish a criminal for his misdeeds as beating up on your car when it breaks down. The criminal cannot do anything about it, for his genetic make-up and his circumstances made him into what he is. He can do no other. The only thing that would help is to repair his brain just as one repairs a computer that is not working properly. If we ever become able to do that we will leave concepts like responsibility and guilt behind us and laugh at them as outrageously primitive views.[3]

Reification

Dawkins' idea that punishing a criminal is just as senseless as kicking a broken-down car is a good example of the reduction of living beings, including humans, to things. This reification, reducing everything to things, is the consequence of the fact that we think we only really understand something due to our left brain, when we can compare it with something we can make ourselves – as if the world was put together by mechanics.

Therefore, as long as science is identified with materialism or physicalism, we have a serious problem, which could also be formulated as follows: how can the reality we encounter in the world have significance? Is significance something we dream up out of our personal perspective, perhaps in order to satisfy our personal subjective needs? Or is there also objective significance, objective values in the world? These would only then be valid if others, just like me, were feeling and autonomously acting individuals who

17. AFTERWORD: IS THERE AN ALTERNATIVE TO MATERIALISM?

experience the world, and not predetermined zombies. But such a notion is not exactly supported by materialism. Dawkins, Lamme and Dennett (to a certain extent) are quite clear about it: we are automatons. Does this view help us in any way?

Behaviourism

This becomes a particularly distressing question when we take a look at the history of behaviourism, which according to Victor Lamme is the only scientific form of psychology. In behaviourism only the perspective of the third person is taken seriously; all inner experiences and deliberations are set aside as mental illusions.

The beginning of experimental behaviouristic psychology is characterised by unethical – to say the least – and outright cruel experiments. John Broadus Watson, the founder of the school of behaviouristic psychology, was inspired by the conditioning experiments of Ivan Petrovitch Pavlov with dogs. Watson did an experiment with a baby of almost nine months, Little Albert. He let the little boy play with a rat, a white rabbit, a dog and a monkey, which the little boy did eagerly and without any fear. He then put the boy on a mattress and gave him the rat. At the same time, he violently struck a piece of steel with a hammer behind the boy's back. The child started to cry. When this scene had been repeated a couple of times Albert became afraid of the rat, also without the sound. Convinced that through conditioning one could make or break a child, Watson said:

> Give me a dozen healthy infants, well-formed, and my own specified world in which to bring them up in and I'll guarantee to take any one at random and train him to become any type of specialist I might select – doctor, lawyer, artist, merchant chief, and, yes, even beggarman and thief, regardless of his talents, penchants, tendencies, abilities, vocations, and race of his ancestors. [4]

In the BBC television programme, *Focus*, presented by physician Michael Mosley, a video was shown of this 'Little Albert experiment.' An effort to find out what happened later to this boy ended somewhat sadly at his little grave. Whether his early death had to do with the experiments was not made clear.

Other atrocious experiments were done in those days. Pavlov, for instance, not only experimented with dogs, but also with children. In order to measure their reaction to seeing food, and later to hearing the bell announcing the meal, he operated on them in such a way that, just as with the dogs, their saliva ran out through a hole in the bottom of their mouths. That too was recorded on film.

Ethics

Now we might think that these horrific stories – and Mosley showed many more than these – are errors from the beginnings of psychological research, and that the application of ethical rules has put a stop to this kind of inhumanity. But what sort of ethics do we then invoke? Neurodeterminism has nothing to offer in this regard.

Ethics, for sure, is impossible without shared values. When I do not view others as autonomous and responsibly acting beings, and do not need to take seriously what others experience, strive for and think, the only source from which I can derive my values is myself. In that case, these values can be no other than: 'what is good for me is good.' In fact, this is the sole motive that neurodeterminists take seriously. Neuroscientist Martin de Munnik, a coworker of Victor Lamme, said in a newspaper interview that in brain scans we cannot find a spot for altruism: 'Altruism proves to be a fiction; it cannot be localised in the brain. There is only reciprocity: I do something for you if you do something for me.'[5] Frans de Waal, on the other hand, shows many examples of cross-species altruism in the animal world.[6] And animals do not expect anything in return for their help. Would human beings really be inferior to them?

In his book *The Possibility of Altruism*, philosopher Thomas Nagel sees what he calls *barter* as the consequence of this view of altruism:

someone who considers another person solely as a means toward realisation of his own objectives will have to take into account that the other will view him in the same way. Such a person will find himself in great personal loneliness, as the only human being in the world surrounded by more or less useful things, some which are able to talk and can be obstacles. He calls it *practical solipsism* (Latin *solus-ipse*: only self)[7] It is the description of the world view of the psychopath. The neurodeterminist has no better alternative to offer. Needless to say that I do not accuse neurodeterminists of immorality. It is quite possible that some of them are people of high moral standing. But materialism has no insights to offer them in this regard.

Fortunately, it is abundantly clear that we are not automatons. Automatons do not develop themselves. One of the unique aspects of being a general physician is that you can experience people through a large part of their lives. I really got to know some of my patients during the full thirty-seven years that I practised. It is inspiring to be able to follow their development. For instance, I knew difficult and unhappy children who as adolescents completely lost their way, and as adults found their own unique place, so that it was a pleasure to see them in my office. That does not fit with Dawkins' letter to the editor.

The Dead End Alley of Neurodeterminism

The world view of materialism has nothing to offer about the part of reality that has significance and value. All of that is mere illusion. It signifies the end of 'the good, the true and the beautiful.' Love is thus reduced to sexual attraction (which in itself is of course also very interesting), and this is brought back to hormones, neurotransmitters, evolutionary signs of fertility and pheromones, behind which the only thing that appears to exist is the urge to spread our selfish genes. The fact that materialism knows nothing about significance and value is perfectly evident in our culture of which it is no coincidence that it is called materialistic. In such a culture something only has value if the consumer thinks it has money value. As Oscar Wilde once remarked when speaking of cynicism: a materialist is someone who knows the price of everything and the value of nothing.

Daniel Dennett, the generally accepted conscience of neurodeterminists, disapproves of every effort to explain consciousness or, even worse, the *self*, as non-materialistic. For then we are 'opening the door to the supernatural.' But the only reason why that would be a 'danger' is because of the old agreement that only matter forms part of nature. Within materialism life would, strictly speaking, also be supernatural. After all, we are still not able to make life in the laboratory out of lifeless substances. Apparently, materialism is incapable of faithfully describing the whole of nature.

Materialism is therefore a dead-end alley as long as we do not add a dimension to it that recognises *life*, *consciousness* and *self* as non-material phenomena. Matter cannot live and die, it cannot wake and sleep, it cannot experience and will anything. Materialists say that once matter has become sufficiently complex in its organisation, it will be able to do those things. We just have to wait a while for the explanation and description of the manner in which this process will take its course. That kind of thing is called proof on credit.

Then what?
There are people who think that quantum physics has answers here, because it introduced non-locality, the influence of the observer on the observed, and uncertainty over against the determinism of physics. These are concepts that can be applied to consciousness so that materialism can be saved. Mathematician Sir Roger Penrose and anesthesiologist Stuart Hameroff have been working on such an explanation for years. They think that quantum phenomena in the microtubuli (tiny tubes of nano size in the cell) of neurons are the carriers of consciousness. In a lecture in Amsterdam, however, they came out with the far from neurodeterministic view that the brain is clearly not a computer, and consciousness is everywhere present.

17. AFTERWORD: IS THERE AN ALTERNATIVE TO MATERIALISM?

In this context a quotation from Erwin Schrödinger, the discoverer of the wave equation in quantum mechanics is applicable:

> I am very astonished that the scientific picture of the real world around me is deficient. It gives a lot of factual information, puts all our experience in a magnificently consistent order, but it is ghastly silent about all and sundry that is really near to our heart, that really matters to us. It cannot tell us a word about red and blue, bitter and sweet, physical pain and physical delight; it knows nothing of beautiful and ugly, good or bad, God and eternity. Science sometimes pretends to answer questions in these domains, but the answers are very often so silly that we are not inclined to take them seriously.[8]

The Phenomena Themselves

But if materialism does not suffice, what will? Many people, including many scientists, are convinced that they do not have a philosophy as the basis of their way of thinking, no fundamental conviction. They appear to base themselves on materialism or physicalism without realising that *that*, too, is a philosophy of life, or a fundamental conviction. The great advantage of physicalism is that it can be immediately understood by anyone. Perhaps this is the reason why to many people it does not look like a philosophy or view of life.

Some people even think that materialism is an *attainment* of 'modern science,' and that therefore once and for all it possesses the claim of being the sole general truth. As we have seen, the truth is a bit different: materialism is no outcome of research, it is an *agreement*, a *guideline* for research into nature. And, not surprisingly, input determines output.

A science that wants to make progress in exploring *life*, *consciousness* and *self* has to admit more than matter. But how? What philosophical foundation can play a role in this? Fortunately a number of ideas have already been formulated about this. For instance, in Chapter 6 we quoted a number of philosophers who are not troubled by the materialistic prejudice. They all take the position that consciousness is not localised in the head. They take the phenomenon seriously that a tree is where we see it, not in a picture in the back of our head. *Phenomenology*, for example, is a research method that bases itself, among other things, on that sober observation.

A number of more philosophically oriented neuroscientists, such as Vittorio Gallese, the co-discoverer of mirror neurons, neurophilosopher Alva Noë and Iain McGilchrist, the psychiatrist who so extensively mapped the roles of both brain hemispheres, and psychiatry professor Thomas Fuchs, declare themselves indebted to phenomenology. They like to cite Maurice Merleau-Ponty, one of the philosophers who want to re-introduce the first-person point of view into thinking. Phenomenologists Merleau-Ponty, Franz Brentano and Edmund Husserl criticised the abstract descriptions of reality in the philosophy of their day; they wanted to go back to the 'phenomena themselves,' the way they actually present themselves.

Experience then becomes important again. For it is from our own experience that we know much about life, consciousness and self that does not come to us via the detour of the third-person point of view as, for instance, is demanded by materialism. By the way, the reluctance to accept the first-person point of view is completely understandable, for it creates the potential for prejudices and emotions that compromise observation. However, materialism also produces its own prejudices. Merleau-Ponty shows how both types of prejudices can be avoided. In the foreword to his book *Phenomenology of Perception* he writes:

> What is important is that we need to describe, not explain, nor analyse ... we need to return to the things themselves, which means in the first place to leave science behind. I am not the result or the crossing point of various causes that determine my body or my psyche, I cannot view myself merely as a part of the world, as the simple object of biology, psychology and sociology.

17. AFTERWORD: IS THERE AN ALTERNATIVE TO MATERIALISM?

> Nor can I shut myself out of the world of science. Everything I know of the world, even with the help of science, I know out of my own point of view and my own experience, without which the symbols of science would remain meaningless.[9]

That is of course precisely the problem. Even when we want to strive for third-person objectivity, we, as the ones who interpret, remain dependent on our own first-person point of view. This can simply never be denied. We see everything, even literally, always out of ourselves. In phenomenological research one has to school oneself in holding back one's own prejudices as much as possible and, as suggested by Merleau-Ponty, 'letting things speak for themselves.' Or one tries to identify with the other, live into the other, in order to discover its significance in that way.

Since the discovery of mirror neurons we may take this step seriously also from a scientific point of view. We may summarise Merleau-Ponty's thinking as follows: a preference for experience over concepts, an emphasis on context, on significance, and on the fundamental role played by the embodied *self* in the living body as the condition for being in the world. For the living (!) body is the medium through which 'intersubjective' experience can be gained.

Psychiatry and Medicine

One profession that openly relies on a phenomenological approach is psychiatry. When psychiatry is reduced to brain processes, the characteristics of the ways in which a variety of psychiatric disorders express themselves become incomprehensible. For example, there is no neuroscientific way to explain the content of a psychosis. This is the reason why neuroscience has so far failed to make any contribution to the classification of psychiatric disorders, as organised in the American diagnostic manual that is used worldwide, *Diagnostic and Statistical Manual of Mental Disorders* (DSM). In the Dutch *Manual for Psychiatry and Philosophy* more than a quarter of its pages are devoted to the phenomenology of psychopathology.[10]

And also in somatic medicine, finally, at long last, the stories of the patients themselves are being considered again, how they experience their illness. After all, illness is for a patient mostly experience. For a couple of decades this was virtually ignored. Attention was focused mostly on how we could remain within the level numbers of the laboratory – cholesterol, sugar, blood pressure, etc. But the real crucial values, the patients' experiences of what they are going through, are coming back into the profession as 'narrative medicine.'[11] This too is a form of phenomenology of which neurologist Oliver Sacks proved to be a great champion.

Phenomenological Research?

The quotation from Merleau-Ponty might have created the impression that phenomenology consists only of descriptions and not of research. This is completely incorrect. In fact, the first phenomenological researcher was already at work long before this philosophy received its name. This was Johann Wolfgang von Goethe, who did his research in nature from the very same points of view mentioned above.[12] For a long time Goethe was not recognised as a researcher. Who would expect a romantic man of letters to do serious research? Currently, though, we are beginning to see a change in the way his scientific work is viewed.

In actual fact, before the narrowing of scientific interest to the DNA molecule, virtually all research in biology was phenomenological. Growth patterns (such as in embryology) were investigated, for instance, by anatomist Bolk who thus discovered striking similarities in form between humans and newborn mammals as a way of holding back the growth that makes the various animals so different. But today also there are scientists who build their cases out of the phenomena. Paleobiologists are able to reconstruct an entire organism out of a fossilised bone fragment. Stephen Gould and Niles Eldredge discovered by comparing

17. AFTERWORD: IS THERE AN ALTERNATIVE TO MATERIALISM?

forms that evolution does not proceed gradually but makes jumps ('punctuated equilibrium').

But of course we are looking for a current example of the kind of science that can generate this. That is not so difficult. In Chapter 9, I mentioned an example, namely the work of Rosslenbroich and my discussion of it relating to the nervous system. His observation that life and evolution show a striving for autonomy cannot be made if we do not look at the phenomena – the animals themselves, their behaviour and anatomy – but only at DNA, as this has become fashionable under the influence of materialism. When we look at the phenomena, the observation of an increase in autonomy becomes abundantly evident for anyone; it is not an individually subjective idea, but an inter-subjective one. All we need to do is push the microscope aside for a moment and look at the whole. We then witness how a value emerges: autonomy, freedom. And it also becomes clear that this value is introduced by conscious organisms. It is an idea that is embodied by animals and humans. Lifeless nature knows no freedom. That is why it is so difficult for neurodeterminists to acknowledge freedom; for them there is no difference between lifeless and living nature.

It is interesting to set this alongside the book *Freedom Evolves* by physicalist (or *naturalist* as he styles himself) philosopher Daniel Dennett.[13] The book's title holds the same promise as Rosslenbroich, but no animal is ever mentioned in the text. It is only about cells, DNA, mechanisms, decision machines, the human 'user interface' and what these mechanisms and machines produce as seen from a Darwinian point of view – spreading of our genes of course. The interesting thing though is that he does not completely refute common sense (which makes him such an interesting author) and takes for granted that freedom and free will do indeed exist. How else can we explain our culture and our self-evident confidence that we are all responsible for our deeds? (And Dennett is an American and, of course freedom is one of the supreme values all Americans share).

According to Dennett true free will exists only in the human being, and in that we agree with him of course. For we do not hold animals responsible for their deeds, although apes certainly know a form of moral practice, as demonstrated by primatologist Frans de Waal. His books, in which he discusses the behaviour of animals by creeping into their skin, as it were, and by placing it in a broader perspective, are good examples of the power of phenomenology.

But despite the title of his book, *Freedom Evolves* Dennett shows no interest in a biological sense in the manner in which freedom evolved. That is strange for a declared Darwinist, for Darwinism demands that we can be fully explained out of evolution. For Dennett, freedom only begins with the human being, and it is explained by language, which has allowed culture to make a huge jump.

The Advantage of Two Brain Hemispheres

Physicalism/materialism is a favorite of the *left* brain. It delivers useful knowledge of the material, quantitative side of reality, knowledge with which we can manipulate the world. The addition of a phenomenological point of view does not in first instance result in 'power,' but increased understanding that has more to do with quality and value.

The research method of materialism is the reduction of everything to its component parts. Stephen Gould, however, wrote in a newspaper article towards the end of his life that, since we now know that the human being only has one and a half times the number of genes of a roundworm – which would indicate that the organism itself determines what happens with the genes – he had found that the smallest unit of life is the entire organism: 'Organisms must therefore be explained as organisms and not as a combination of genes.' Did he realise that he breathed new life into vitalism with this statement?

Dennett is not the only one to describe an organism as a collaboration of molecules and cells, instead of a whole that is subdivided into cells. That is actually the chief reason for him to view living beings as machines. All cells are some kind of robots. The robots that are neurons all by themselves make consciousness and the illusion of the *self*.

But we are not built up out of cells. We began as one cell that then started to divide. From the beginning, therefore, we have been *one whole* that is subdivided into cells. Not until one considers the whole of an organism, instead of merely the parts as building blocks, does it become possible to recognise meaning and significance, such as the evolutionary striving for freedom. Organisms strive, genes don't. *Significance* becomes visible only when we can see the whole, and that is the domain of the *right* brain.

This is also true for attaching value. A concept like the intrinsic value or integrity of an animal is derived from the right brain. The fact that people object when they hear that chimpanzees are made to suffer in laboratories is incomprehensible out of the mechanistic world view that currently dominates science. As a matter of fact, science will advance all kinds of arguments to prove how useful the experiments on chimpanzees are.

In Balance

The point is not to trade in materialistic science with its evident benefits for phenomenological research. The latter serves entirely different purposes: not power or manipulation, but insight and understanding of what we are and what the world is, by learning to see the significance of things. We may quarrel about what is more important, quantitative or qualitative understanding of the world, but as Einstein is supposed to have said: *not everything that can be counted counts; and not everything that counts can be counted.*

Science without the preconception that matter is the only thing that exists has the potential to approach things that are clearly not physical, such as consciousness, without prejudice. It won't be afraid of 'opening the door to the supernatural.' It can also return to consciousness the importance that it lacks in neurodeterministic theories (illusion, epiphenomenon).

Materialism, the idea that all of reality has to be explicable in terms of physics originated in the conviction that physics is the mother of all sciences. But today it is precisely the physicists who are no longer afraid of a non-physical reality. The superstring theory presumes eleven dimensions, ten in space and one in time, and its 'Braneworld' model is based on the presumption that matter occurs only in the three-dimensional world we can perceive; the other dimensions are non-material.

By the way, I do not want to suggest that this model explains consciousness, supposing I understand it at all. Fortunately, the idea of understanding is not really applicable here; the model can be computed and could address some unsolved problems of quantum mechanics regarding gravity.

Reification (Again)

Author Rudy Kousbroek one day thought he had found the solution to all the misery in the world: 'Scientists of all countries, unite!' Today's 'value-free' science has been very successful and has brought us much good but, because of its lack of values, it has also saddled us with a number of unpalatable problems. Think of environmental pollution, possible climate change, 'prosperity illnesses' due to the food industry, genetic modification of plants to permit unlimited spraying with weed killers, medicines which retrospectively turned out to make more victims than healing, resistance to antibiotics, the dying of bees, the atom bomb. And this list is of course far from exhaustive.

In the past it was not any better; eugenics in the sense of sterilising 'undesirable elements' was common in most western countries before World War II, and Nazism was based on Darwinism. Experiments on prisoners also were not limited to the Nazis. These things are explained not as a direct consequence of science, but as the responsibility of the people making use of it. Can we really make this distinction? No reproach can ever be made of science, for it has no will; it is the scientists who are human all too human in their desires and passions. When a scientific world view leads to reification, to turning living beings into things, problems like those mentioned here are to be expected.

17. AFTERWORD: IS THERE AN ALTERNATIVE TO MATERIALISM?

How can we rid ourselves of reification? Very simply by ceasing to insist at all cost that the only thing that exists is matter, just as Benjamin Libet concluded. Consciousness is not physical; it helps us understand physics. Science made immense progress in a time – which lies not very far behind us – when everyone was familiar with the concepts of soul and spirit. Soul and spirit formed part of nature. Maybe the problem began when spiritualistic charlatans started to make photographs of the soul as 'ectoplasm,' together with ghost apparitions. They were actually the first ones who viewed the spirit as material.

For that matter, we don't want to trade the left brain for the right one either, even if it were possible. How can we see to it that the right brain regains the useful place that is its due? In the preface to the Dutch translation of the book *The Social Brain*[14] by Michael Gazzaniga, Dutch psychologist Piet Vroon writes that in order to stimulate the right brain we should breathe through our left nostril. Would that it were that easy! Then we could advise neurodeterminists to visit their ear-nose-and-throat doctor to open up their left nostril.

When we follow the line of thought of McGilchrist in Chapter 13, the right brain is put into action by visiting, observing and appreciating nature, as we find it out in the open, not in the form of detecting DNA (which may be useful in itself) but as whole organisms, living beings. It is of course no coincidence that many of us go out into nature in our vacations in order to recover from the rat race driven by the left brain. Teaching nature out in the field is something that should start already in grade school, just like visual arts and music. These things stimulate the right brain and therefore they ought to be continued during our entire life. The same holds true for reading novels (also from other cultures), biographies (which can help develop empathy) and poetry, as well as the practice and appreciation of other art forms. Our man-made surroundings, which in their current form stimulate the left brain, might be given a rather more organic form. There are forms that are more liveable than our typical straight lines and square corners (which, I admit, can also have beauty). And we might try for ourselves to silence the chatterbox of our left brain, for instance, by meditation. According to a meta-analysis of a very large number of studies of the psychological effects of meditation,[15] the result of this is that we suffer less from negative emotions and anxieties, and can develop better relationships with people. All of these are efforts that foster not only our spiritual but also our physical health. The two go together.

And why should we bring both brain hemispheres in balance again? Why should we broaden science with phenomenology? For at least two reasons. First, because otherwise significance and value will vanish, so that we can no longer give an answer to the question of what makes the world and a human life important and worthwhile. And second, in order to prevent the field of science from being closed any longer to an important part of reality.

We need both hemispheres of the brain just as we need two eyes, namely to perceive depth. Can anyone think of a better reason?

Notes

Inside cover
- A   Nummenmaa, L. *et al.*, *Bodily Maps of Emotions*, PNAS, 30 December 2013
- B   www.what-when-how.com/neuroscience/overview-of-the-central-nervous-system-gross-anatomy-of-the-brain-part-1.
- C   The evolutionary layers of the human brain: www.thebrain.mcgill.ca/flash/i/i_05/i_05_cr/i_05_cr_her/i_05_cr_her.html.
- E   University of Southampton Blogs: www.blog.soton.ac.uk/comp6044/files/2010/11/09-15.jpg.

Figures
1. Perry B. D. & Pollard, R., *Altered brain development following global neglect in early childhood*, Society for Neuroscience: Proceedings from Annual Meeting, New Orleans, 1997.
2. Posner, M. I. & Raichle, M. E., *Images of Mind*, The Scientific American Library, New York, 1994.
4. *Brainport remaps sensory input to tongue*: www.geekdoc.org/wordpress/tag/hardware.
6. www.brainmind.com/ParietalLobes.html.
7. www.neurowiki2012.wikispaces.com/Phantom+Sensations+Perceptions.
9. Bio1152.nicerweb.com/Locked/media/ch49/49_02aNervousSystemsA-U.jpg.
10. Toga, A. W. & Thompson, P. M., *Mapping Brain Asymmetry*, in: Nature Reviews, Neuroscience 4: 37-48, 2003.
11. www.plato.stanford.edu/entries/mental-imagery/representational-neglect.html.
13. Feuillet, L., *Brain of a White-Collar Worker*, in: Lancet 370; 9583; 262, 2007.
14. James Fallon photo: www.faculty.uci.edu/profile.cfm?faculty_id=2303.

Foreword
1. Swaab, D., *We Are Our Brains: From the Womb to Alzheimer's*. Allan Lane, London 2014.
2. Cutting, G., 'What Do Scientific Studies Show?' Opinion Pages *The New York Times* online April 25, 2013.

Chapter 1
1. Such as: Dennett, D., *Consciousness explained*, Little, Brown & Co 1991; Pinker, S., *The Blank Slate. The Modern denial of Human nature*, Penguin Putman 2002; Pinker, S., *How the Mind Works*, Norton & Co, New York 1997; Tiger L., and Mcguire, M., *God's Brain*, Prometheus Books 2010. I will use the term 'brain books' for those books that are based on the thesis that the brain produces our consciousness.

2   Crick, F.H.C., *The Astonishing Hypothesis. The Scientific Search for the Soul*, Simon & Schuster, London 1994.
3   Dijksterhuis, A., *Het slimme onbewuste: Denken met gevoel* [The smart unconscious: thinking with feeling], Bert Bakker, Amsterdam 2008.
4   Gerhardt, S., *Why Love Matters. How Affection Shapes a Baby's Brain*, Routledge, Oxford 2004.
5   Bick J., et al., 'Effect of Early Institutionalization and Foster Care on Long-term White Matter Development. A Randomized Clinical Trial,' *JAMA Pediatrics* 196 (3): 211–219. 2015.
6   Bos, A., *Hoe de stof de geest kreeg* [How matter got spirit], Christofoor, Zeist 2010.
7   Anders Ericsson, K., *The Road to Excellence. The Acquisition of Expert Performance in the Arts and Sciences, Sports and Games*, Lawrence Erlbaum Ass, New Jersey 1996.
8   Nuzo, R., 'Babies' Brains May Be Tuned to Language before Birth,' *Nature* doi:10.1038/nature.2013.12489.
9   Borgstein, J. & Grootendorst, C., 'Half a Brain,' *The Lancet* 2002: 359, 9305, 473.
10 Tarr, M.J. & Gauthier,I., 'FFA: A Flexible Fusiform Area for Subordinate-Level Visual Processing Automatized by Expertise,' *Nature Neuroscience*, 3, 8: 764–769, 2000.
11 Tan, C.B.Y., 'You Look Familiar: How Malaysian-Chinese Recognize Faces,' *PLoS ONE* 7,1: e29714, 2012.
12 Pascalis, O., et al., 'Is Face Processing Species-Specific during the First Year of Life?' *Science* 296: 1321–1323, 2002.
13 Scott, L.S. & Monesson, A., 'The Origin of Biases in Face Perception,' *Psychological Science* 20: 676–680, 2009.
14 Rossion, B., et al., 'A Network of Occipito-Temporal Face-Sensitive Areas Besides the Right Middle Fusiform Gyrus is Necessary for Normal Face Processing,' *Brain* 126 (11): 2381–2395, 2003.
15 Noë, A., *Out of Our Heads. Why You Are Not Your Brain, and Other Lessons from the Biology of Consciousness*, Hill & Wang, New York 2009.
16 Bernstein, M.J., 'The Cross-Category Effect. Mere Social Categorization Is Sufficient to Elicit an Own-Group Bias in Face Recognition,' *Psychological Science* 18, 8: 706–712, 2007.
17 Aarsman, H., *De fotodetective*, Podium, Amsterdam 2012.
18 Heinrich, J., Heine, S.J. & Norenzayan, A., 'The Weirdest People in the World?' *Behavioral and Brain Sciences* 33, 2/3: 61–135, 2010.
19 Segall, M.H., Campbell, D.T. & Herskovits, M.J., *The Influence of Culture on Visual Perception*, Bobbs-Merrill, Indianapolis 1966.
20 Lotto, R.B., 'Visual Development: Experience Puts the Colour in Life,' *Current Biology* 14: R619–R621, 2004.
21 Guo, J. U., et al., 'Neuronal Activity Modifies the DNA Methylation Landscape in the Adult Brain,' *Nature Neuroscience* 14: 1345–1351, 2011.
22 Recently neurogenesis has also been found in the striatum (basic nuclei such as nucleus caudatus, putamen and nucleus accumbens. Welberg, L., 'A Striatal Supply of New Neurons,' *Nature Reviews Neuroscience* 15; 203, online March 5, 2014.

23 Coufal, N.G., Garcia-Perez, J.L., Gage, F.H., et al., 'L1 Retrotransposition in Human Neural Progenitor Cells,' *Nature* 460: 1127–1131, 2009.
24 Nisbett, R.E., Flynn, W., et al., 'Intelligence: New Findings and Theoretical Developments,' *American Psychologist*, January 2, 2012 (online).
25 Cole, M., et al., 'A Smart Hub in the Brain,' *Nature* 487, 7407: 275, July 19, 2012.
26 Covic, M., et al., 'Epigenetic Regulation of Neurogenesis in the Adult Hippocampus,' *Heredity* 105: 122–131, 2010. Dragnski, B., et al., 'Temporal and Spatial Dynamics of Brain Structure Changes during Extensive Learning,' *Journal of Neuroscience* 26, 23: 6314–6317, 2006.
27 LeDoux, J., *Synaptic Self: How Our Brains Become Who We Are*, Viking Penguin, 2002.

Chapter 2

1 Aguilar, M.J., 'Recovery of Motor Function after Unilateral Infarction of the Basis Pontis; Record of a Case,' *American Journal of Physical Medicine and Rehabilitation* 48, 6:279–288, December 1969.
2 Doidge, N., *The Brain That Changes Itself. Stories of Personal Triumph from the Frontiers of Brain Science*, Penguin, London 2007.
3 Amedi, A., et al., 'The Occipital Cortex in the Blind. Lessons about Plasticity and Vision,' *Current Directions in Psychological Science* 14: 6. 2005.
4 Khamsi, R., 'Plastic Brains Help the Blind Place Sounds. Brain Areas for Vision Are Co-Opted for Locating Sound in Space,' News@nature.com, January 25, 2005.
5 Bedny, M., Caramazza, A., et al., 'Typical Neural Representations of Action Verbs Develop without Vision,' *Cerebral Cortex* June 7, 2011 (online, in print).
6 Kendrick, M., 'Tasting the Light,' *Scientific American* 301: 22–24, 2009.

Chapter 3

1 Hoffer, E., *The True Believer. Thoughts on the Nature of Mass Movements*, Harper & Row, New York 1951.
2 Nietzsche, F., *Morgenröthe, Gedanken über die moralische Vorurteile*, in Nietzsche, F., *Werke*, Naumann Leipzig 1899, vol. 2, p. 142. Quoted in Fuchs, T., *Das Gehirn: ein Beziehungsorgan. Eine phaenomenologisch-ökologische Konzeption*, Kohlhammer, Stuttgart 2010.
3 Ramachandran, V.S., *The Tell-Tale Brain. A Neuroscientific Quest for What Makes Us Human*, W. W. Norton & Company, New York/London, 2011.
4 Gallese, F., Fadiga, L., Fogassi, L. & Rizolatti, G., 'Action Recognition in the Premotor Cortex,' *Brain* 119: 593–609, 1996. And Rizolatti, G., Fadiga, L. & Gallese V., 'Premotor Cortex and the Recognition of Premotor Actions,' *Cog. Brain Res.* 3: 131–141, 1996.
5 Mukamel, R., Ekstrom, A.D., Kaplan, J., Iacoboni, M. & Fried, I., 'Single-Neuron Responses in Humans during Execution and Observation of Actions,' *Current Biology* 20: 750–756, 2010.
6 Ibid.
7 Can be viewed on www.youtube.com as 'Merci! by Christine Rabette (Bodhisattva on Metro)'.

8   This example is based on a test of people's appreciation of cartoons, with or without having a pencil clenched between their teeth. It was found that people liked the cartoons better when they had a pencil in the mouth. Recently an attempt was made to repeat this test, without the same result. There are also many other tests that demonstrate the influence of body language on mood.
9   Pfeiffer, J., et al., 'Mirroring Others' Emotions Relates to Empathy and Interpersonal Competence in Children,' *Neuro Image* 39: 2076–2085, 2008.
10 Ritvo, S. & Provence, S., 'From Perception and Imitation in Some Autistic Children: Diagnostic Findings and their Contextual Interpretation,' *The Psychoanalytical Study of the Child*, Vol. VIII, International Universities Press, New York 1953.
11 Iacoboni, M., *Mirroring People: The New Science of How We Connect With Others*, Farrar, Straus & Giroux, New York 2008.
12 Can be viewed on www.youtube.com as 'Dog tries to save another dog on highway'.
13 De Waal, F. *The Bonobo and the Atheist: In Search of Humanism Among the Primates*, W.W. Norton & Co, New York 2013.
14 Prather, J.F., et al., *Precise Auditory-Vocal Mirroring in Neurons for Learned Vocal Communication*, in: Nature 451: 305–310, 2008.
15 Stenekes, M., *Cerebral Reorganisation and Motor Imaging after Flexor Tendon Injury*, Doctoral Thesis RUG 2009.
16 Buccino, G., et al., 'Functions of the Mirror Neuron System: Implications for Neurohabilitation,' *Cognitive and Behavioral Neurology* 19. 1:55–63. 2006.
17 Keysers, C., *Het empathische brein. Waarom we socialer zijn dan we denken* [The empathetic brain. Why we are more social than we think], Bert Bakker, Amsterdam 2012.
18 Iacoboni, M., *Mirroring People: The New Science of How We Connect With Others*, Farrar, Straus & Giroux, New York 2008.
19 Calvo-Merino, B., 'Action Observation and Acquired Motor Skills: an fMRI Study with Expert Dancers,' *Cerebral Cortex* 15: 1243–1249, 2005.
20 Jabbi, M., Bastiaansen, J. & Keysers, C., 'A Common Anterior Insula Representation of Disgust Observation, Experience and Imagination Shows Divergent Functional Connectivity Pathways,' *PLoS ONE* 3 (8): e2939, 2008.
21 Velzen, T. van, 'Magneten tegen autisme' [Magnets against autism], *NWT Magazine* 80, 1: 33–37, 2012.
22 Fein, D., et al., 'Optimal Outcome in Individuals with a History of Autism,' *Journal of Child Psychology and Psychiatry* 54, 2: 195–205, 2013.
23 Vandermassen, G., 'Bioloog en psycholoog Christian Keysers: Spiegelneuronen laten zien hoe door en door sociaal wij zijn' [Mirror neurons show how thoroughly social we are], *Psyche & Brein* 4, 2012.
24 Meffert, H., et al., 'Reduced Spontaneous but Relatively Normal Deliberate Vicarious Representations in Psychopathy,' *Brain*, 136, 8: 2550–2562, 2013.

Chapter 4

1   Waldvogel, B., Ulrich, A. & Strasburger, H., 'Blind und sehend in einer Person. Schlussfolgerungen zur Psychoneurobiologie des Sehens' [Blind and sighted in

one person: conclusions of the psycho-neurobilogy of sight], *Nervenartzt* 78: 1303–1309, 2007 (online July 2007).
2. Baxter, R.L., Schwartz, J.M., et al., 'Caudate Glucose Metabolic Rate Changes with Both Drug and Behavior Therapy for Obsessive Compulsive Disorder,' *Archives of General Psychiatry* 49 (9): 681–689, 1992.
3. Brody, A.L., et al., 'Regional Brain Metabolic Changes in Patients with Major Depression Treated with Either Paroxetine or Interpersonal Therapy: Preliminary Findings,' *Archives of General Psychiatry* 58 (7): 631–640, 2001.
4. Engel, A.K., 'Neuronale Grundlagen der Merkmalsintegration,' Chapter 5 of *Neuropsychologie*, Springer, Heidelberg 2006.
5. Hopf, J.M., et al., 'Localizing Visual Discrimination Processes in Time and Space,' *Journal of Neurophysiology* 88: 2088–2095, 2002.
6. Cerf M., et al. 'On-line, voluntary control of human temporal lobe neurons,' *Nature* 467, 1104–1108. 2010.
7. Jones, R., 'Development: A Light Touch for Eye Development,' *Nature Reviews Neuroscience* 14, 156–157, 2013.
8. Haldane, J.B.S., 'When I Am Dead,' in *Possible Worlds and Other Essays,* Chatto & Windus, London 1932.
9. Fuchs, T., 'The Brain: A Mediating Organ,' *Journal of Consciousness Studies* 18, 7–8: 196–221, 2011.

Chapter 5

1. Eeden, F. van, 'Ons Dubbel-ik' [Our double I], p. 177 in *Studies. Eerste reeks*, W. Versluys, Amsterdam 1891.
2. Weisberg, D.S., et al., 'The Seductive Allure of Neuroscience Explanations,' *Journal of Cognitive Neuroscience* 20 (3): 470–477, 2008.
3. McCabe, D.P. & Castel, A.D., 'Seeing is Believing: The Effect of Brain Images on Judgments of Scientific Reasoning,' *Cognition* 107: 343–352, 2008.
4. Lamme, V., *De vrije wil bestaat niet. Over wie er echt baas is in het brein* [Free will does not exist: about who is really boss in the brain], Bert Bakker, Amsterdam 2010.
5. Ganis, G., et al., 'Lying in the Scanner: Covert Countermeasures Disrupt Deception by Functional Magnetic Resonance Imaging' *Neuroimage* 55: 312–319, 2011.
6. Eeden, F. van, 'Ons Dubbel-ik' [Our double I], p. 177 in *Studies. Eerste reeks*, W. Versluys, Amsterdam 1891.
7. Bennett, C.M., et al., *Neural Correlates of Interspecies Perspective Taking in the Post-Mortem Atlantic Salmon: An Argument for Multiple Comparisons Correction*, Poster at the Human Brain Mapping Conference, June 2009, San Francisco.
8. Can be viewed on www.youtube.com as 'Wie man dem toten Hasen die Bilder erklärt'.
9. Eklund A. et al. 'Cluster failure: Why fMRI inferences for spatial extent have inflated fals-positive rates,' *PNAS* 113, 287900–7905. 2016.
10. Dobbs, D., 'Fact or Phrenology?' *Scientific American Mind* 24–31, April 2005.
11. Kanai, R. & Rees, G., 'Opinion: the Structural Basis of Inter-Individual Differences in Human Behaviour and Cognition' *Nature Reviews Neuroscience* 12: 231–241, 2011.
12. Kong, J., et al., 'Test-Retest Study of fMRI Signal Change Evoked by Electro-Acupuncture Stimulation' *Neuroimage* 34, 3: 1171–1181, 2007.

13. Haasabis, D., et al., 'Imagine All the People: How the Brain Creates and Uses Personality Models to Predict Behaviour,' *Cerebral Cortex* online March 5, 2013
14. Vu, E., et al., 'Puzzling High Correlations in fMRI studies of Emotion, Personality and Social Cognition,' *Perspectives on Psychological Science* 4, 3: 274–290, 2009.
15. Beauregard, M., et al., 'The Neural Basis of Unconditional Love,' *Psychiatry Research Neuroimaging* 172, 2: 93–98, 2009.
16. Kawabata, H. & Zeki, S., 'Neural Correlates of Beauty,' *Journal of Neural Psychology* 91: 1699–1705, 2004.
17. Bartels, A. & Zeki, S., 'The neural Basis of Romantic Love' *Neuroreport* 11, 17: 3829–3834, 2000.
18. Fisher, H., et al., 'Romantic Love: an fMRI Study of a Neural Mechanism for Mate Choice,' *Journal of Comparative Neurology* 493, 1: 58–62, 2005.
19. Tallis, R., *Aping Mankind. Neuromania, Darwinitis, and the Misrepresentation of Humanity*, Acumen, Durham 2011.
20. Parri, R. & Crunelli, V., 'An Astrocyte Bridge from Synapse to Blood Flow,' *Nature Neuroscience* 6: 5–6, 2003.
21. Sirotin, Y.B. & Das, A., 'Anticipatory Haemodynamic Signals in Sensory Cortex Not Predicted by Local Neuronal Activity,' *Nature* 457, 475–479, 2009.

Chapter 6

1. Merleau-Ponty, M., *Phénomenologie de la Perception*, Gallimard, Paris 1945.
2. Bastuji, H., et al., 'Filtering the Reality: Functional Dissociation of Lateral and Medial Pain Systems during Sleep in Humans,' *Human Brain Mapping* 33 (11): 2638–2649, 2012.
3. Libet, B., 'Subjective Referral of the Timing for a Conscious Sensory Experience,' *Brain* 102: 193–224, 1979.
4. Churchland, P., 'On the Alleged Backwards Referral of Experiences and its Relevance to the Mind-Body Problem,' *Philosophy of Science* 48: 165–181, 1981.
5. Libet, B., 'Reflections on the Interactions of the Mind and the Brain,' *Progress in Neurobiology* 78:322–326, 2006.
6. Paul, R.L., Goodman, H. & Merzenich, M.M., 'Alterations in Mechanoreceptor Input to Brodman's Areas 1 and 3 of the Post-Central Hand Area of Macaca Mulatta after Nerve Section and Regeneration,' *Brain Research* 39 (1): 1–9, 1972; and by the same authors, 'Representation of Slowly and Rapidly Adapting Cutaneous Mechanoreceptors of the Hand in Brodman's Areas 3 and 1 of Macaca Mulatta,' *Brain Research* 36 (2): 229–249, 1972.
7. Rivers, W.H.R. & Head, H., 'A Human Experiment in Nerve Division,' *Brain* XXXI: 323–450, 1908 (also en.wikisource.org/wiki/A_human_experiment_in_nerve_division/Chapter_1).
8. Flor, H., 'Maladaptive Plasticity, Memory for Pain and Phantom Limb Pain: Reviews and Suggestions for New Therapies,' *Expert Rev. Neurotherapeutics* 8 (5): 809–818, 2008.
9. Ramachandran, V.S., *The Tell-Tale Brain: A Neuroscientific Quest for What Makes Us Human*, W.W. Norton, New York/London, 2011.
10. Makin, T.R., et al., 'Phantom Pain is Associated with Preserved Structure and Function in the Former Hand Area,' *Nature Communications* 4, Article number 1570, March 5, 2013.

NOTES

11 Jensen, T.S., et al, 'Immediate and Long-Term Phantom Limb Pain in Amputees: Incidence, Clinical Characteristics and Relationship to Pre-Amputation Limb Pain,' *Pain* 21, 3: 267–278, 1985.
12 Maassen, H., & Ophey, P., 'Flexibele hersenen' [Flexible brains], *Wetenschap, Cultuur en Samenleving* July–August: 60–63, 1996.
13 Bos, A., *Hoe de stof de geest kreeg* [How matter got spirit], Christofoor. Zeist 2008.
14 Kelso, J.A.S., et al., 'Functionally specific articulatory cooperation following jaw perturbations during speech: Evidence for coordinative structures,' *Journal for Experimental Psychology: Human Perception and Performance* 10 (6): 812–832, 1984.
15 Wallot, S. & Van Orden, G., 'Ultrafast Cognition,' *Journal of Consciousness Studies* 19, 5–6: 141–160, 2012.
16 Lashley, K.S., *In Search of the Engram, Society of Experimental Biology Symposium, No. 4: Physiological Mechanisms in Animal Behaviour,* Cambridge University Press, Cambridge 1950, pp. 454f, 468–73, 477–80.
17 Pribram, K.R., *Holographic Hypothesis,* Nuwer & Barron, quoted in: Draaisma, D., *De metaforenmachine,* Historische Uitgeverij, Groningen 2010.
18 Bor, D., 'This is Your Brain on Consciousness,' *New Scientist* 218, 2917; 32, 2013.
19 Soltmann, O., 'Experimentelle Studien über die Funktionen des Grosshirns der Neugeborenen,' *Jahrbuch für Kinderheilkunde und physische Erziehung,* 9:106–148, 1876, quoted in Doidge, N., *The Brain that Changes Itself. Stories of Personal Triumph from the Frontiers of Brain Science,* Penguin, London 2007.
20 Sur, M., Angelucci, A., and Sharma, J., 'Rewiring cortex: The Role of Patterned Activity in Development and Plasticity of Neocortical Circuits,' *Journal of Neurobiology* 41, 1: 33–43, 1999.
21 Noe, A., *Out of Our Heads. Why You Are Not Your Brain, and Other Lessons from the Biology of Consciousness,* Hill & Wang, New York, 2009.
22 Mountcastle, V.B., 'Modality and Topographic Properties of Single Neurons of Cat's Somatic Sensory Cortex,' *Journal of Neurophysiology* 20 (4): 408–34, 1957.
23 Campbell, F., Macsweeney, M. and Waters, D., 'Sign Language and the Brain: A Review,' *Journal of Deaf Studies and Deaf Education* 13:1, Winter 2008.
24 Bennett, M.R. & Hacker, P.M.S., *Philosophical Foundations of Neuroscience,* Blackwell, Oxford 2003.
25 Quoted in Sheldrake, R., *The Science Delusion: Freeing the Spirit of Enquiry,* Coronet, London 2012.
26 www.theinvisiblegorilla.com
27 McGilchrist, I., *The Master and His Emissary. The Divided Brain and the Making of the Western World,* Yale University Press 2009.
28 Langer, E., 'Believing is Seeing. Using Mindlessness (Mindfully) to Improve Visual Acuity,' *Psychological Science* 21,5: 660–666, 2010.
29 James, W., 'Does "Consciousness" Exist?' *Journal of Philosophy, Psychology and Scientific Methods,* i: 477–491, 1904.
30 Whitehead, A.N., *Science and the Modern World,* Macmillan, New York 1925, p. 54.
31 Bergson, H., *Creative Evolution,* Macmillan London 1911, quoted in Sheldrake, R., *The Science Delusion. Freeing the Spirit of Enquiry,* Coronet, London 2012.

32 Merleau-Ponty, M., *Phénomenologie de la Perception,* Gallimard, Paris 1945.
33 Noë, A., *Out of Our Heads: Why You Are Not Your Brain, and Other Lessons from the Biology of Consciousness,* Hill & Wang, New York 2009.
34 Fuchs, T., 'The Brain: A Mediating Organ,' *Journal of Consciousness Studies* 18, 7–8: 196–221, 2011.

Chapter 7

1 Swaab, D., 'Geen gedachte zonder fosfor' [No thought without phosphorus], *AMC Magazine,* December 2002.
2 Dennett, D.C., *Consciousness Explained,* Back Bay Books 1992.
3 Putnam, H., *Reason, Truth and History,* Cambridge University Press 1982.
4 Grayling, A.C., 'Psychology: How We Form Beliefs,' *Nature* 474, 446–447, 2011.

Chapter 8

1 Damasio, A., *Descartes' Error: Emotion, Reason, and the Human Brain,* Putnam, 1994; revised Penguin edition, 2005.
2 James, W., 'What is an Emotion?' *Mind* 9: 188–205, 1884. Also: psychclassics.yorku.ca/James/emotion.htm
3 Kross, E., et al., 'Social Rejection Shares Somatosensory Representations with Physical Pain,' *PNAS* 108,15: 6270–6275, 2011.
4 Nummenmaa, L., et al., 'Bodily Maps of Emotions,' *PNAS* December 30, 2013 (epub ahead of print).
5 Kahn, R., *Onze Hersenen: Over de Smalle Grens Tussen Normal en Abnormaal* [Our brain: on the narrow boundary between normal and abnormal], Balans, Amsterdam 2006.
6 Strack, F., et al., 'Inhibiting and Facilitating Conditions of the Human Smile: A Non-Obtrusive Test of the Facial Feedback Hypothesis,' *Journal of Personality and Social Psychology* 54, 5: 768–777, 1988.
7 Hennenlotter, A., Haslinger, B., et al., 'The Link between Facial Feedback and Neural Activity within Central Circuitries of Emotion: New Insights from Botulinum Toxin-Induced Denervation of Frown Muscles,' *Cerebral Cortex* 19, 3: 537–542, 2009.
8 Havas, D., Glenberg, A.M., et al., 'Cosmetic Use of Botulinum Toxin Affects Processing of Emotional Language,' *Psychological Science* 27, 7: 895–900, 2010.
9 Lakoff, G. & Johnson, M., *Metaphors We Live By,* University of Chicago Press 1980; same authors, *Philosophy in the Flesh,* Basic Books, New York 1999.
10 Williams, L.E. & Bargh, J.A., 'Experiencing Physical Warmth Promotes Interpersonal Warmth,' *Science* 322, 5901: 606–607, 2008.
11 Kang, Y, Williams, L.E. & Bargh, J.A., 'Physical Temperature Effects on Trust Behaviour: the Role of the Insula,' *Social Cognitive and Affective Neuroscience* 6,4: 507–515, 2010.
12 Zong, C.B. & Leonardelli, G.J., 'Cold and Lonely: Does Social Exclusion Literally Feel Cold?' *Psychological Science* 19, 9: 838–842, 2008.
13 Ackerman, J.M., Bargh, J.A., et al., 'Incidental Haptic Sensations Influence Social Judgments and Decisions,' *Science* 328, 5986: 1712–1715, 2010.
14 Jostmann, N.B., et al., 'Weight as an Embodiment of Importance,' *Psychological Science* 20, 9: 1169–1174, 2009.

15 Zong, C.B., et al., 'A Good Lamp is the Best Police. Darkness Increases Dishonesty and Self-Interested Behaviour,' *Psychological Science* 21: 311–314, 2010.
16 Liljenquist, K. & Zong, C.B., 'The Smell of Virtue: Clean Scents Promote Reciprocity and Charity,' *Psychological Science* 21: 381–383, 2010.
17 Zong, C.B. & Liljenquist, K., 'Washing Away Your Sins: Threatened Morality and Physical Cleansing,' *Science* 313, 5792: 1451–1452, 2006.
18 Zong, C.B., et al., 'A Clean Self Can Render Harsh Moral Judgment,' *Journal of Experimental Social Psychology* 46: 859–862, 2010.
19 Lee, S. & Schwartz, N., 'Dirty Hands and Dirty Mouths: Embodiment of the Moral-Purity Metaphor is Specific to the Motor Modality Involved in Moral Transgression,' *Psychological Science* 21: 1423–1425, 2010.
20 Seldenrijk, A., 'Depressie, angst en subklinische vaatschade' [Depression, anxiety and subclinical damage to blood vessels], *Hartbulletin* 45,5: 114–117, 2011.
21 Blanker, M., 'Cognitieve gedragstherapie helpt ook bij hart- en vaatziekten' [Cognitive behavioural therapy also helps in diseases of heart and blood vessels], *HuisartsenWetenschap* 54, 7: 404, 2011; also: Bennebroek Evertsz, F. & Haes, H. de, 'Minder terugkeer van cardiovasculaire ziekte na psychotherapie' [Less relapse of cardiovascular illness after psychotherapy], *Nederlands Tijdschrift voor Geneeskunde* 155, 17: 809, 2011.
22 Heesch, F. van, *Inflammation-induced Depression. Studying the Role of Pro-Inflammatory Cytokines in Anhedonia,* Utrecht University, Farmaceutische Wetenschappen Proefschriften 2014; and Brown, P., 'A Mind under Siege,' *New Scientist* 170 (2295): 34–36, 2001.
23 Hong, S., et al., 'The Association between Interleukin-6, Sleep and Demographic Characteristics,' *Brain Behav. Immun.* 19 (2): 165–172, 2005.
24 Vgontzas, A.N., et al., 'Circadian Interleukin-6 Secretion and Quantity and Depth of Sleep,' *Journal of Clin. Endocrinol. Metab.* 84 (8): 2603–2607, 1999.
25 Gerven, M. van & Tellingen, C. van, 'Depressive Disorders: An Integral Psychiatric Approach,' Bolk's Companions for the Practice of Medicine, Driebergen 2010.
26 Ströhle, A., et al., 'Anxiety Modulation by the Heart? Aerobic Exercise and Atrial Natriuretic Peptide,' *Psychoneuroendocrinology* 31,9:1127–1130, 2006.
27 Park, H.D., et al., 'Spontaneous Fluctuations in Neural Responses to Heartbeats Predict Visual Detection,' *Nature Neuroscience* 17: 612–618, 2014.
28 Mayer, E.A., 'Gut Feelings: The Emerging Biology of Gut-Brain Communication,' *Nature Review Neuroscience* 12: 453–466, 2011.
29 Hijmans, H., 'De grote oorlog' [The Great War], *AMC Magazine* January 2012: 8–10.

Chapter 9

1 Grossman, V., *Life and Fate,* New York Review of Books 2006.
2 Perry, G.H., 'The Evolutionary Significance of Copy Number Variation in the Human Genome,' *Cytogenic Genome Research* 123 (1–4): 283–287, 2008.
3 Chen, M., et al., 'Decoupling Epigenetic and Genetic Effects through Systematic Analysis of Gene Position,' *Cell Reports* 3, 1: 128–137, 2013.
4 Nithianantharajah, J., et al., 'Synaptic Scaffold Evolution Generated Components

of Vertebrate Cognitive Complexion,' *Nature Neuroscience* 16: 16–24, 2013; and Ryan, T. J., et al., 'Evolution of GluN2A/B Cytoplasmic Domains Diversified Vertebrate Synaptic Plasticity and Behavior,' *Nature Neuroscience* 16: 25–32, 2013; and the commentary on both articles: 'Retooling Spare Parts: Gene Duplication and Cognition,' *Nature Neuroscience* 16: 6–8, 2013.

5   Dennis, M.I., et al., 'Evolution of Human-Specific Neural SRGAP2 Genes by Incomplete Segmental Duplication,' *Cell* 149: 912–922, 2012.
6   Bauer, J., *Das kooperative Gen. Evolution als kreativer Prozess*, Heyne, Munich 2010.
7   Perry, G.H., et al., 'Diet and the Evolution of Human Amylase Gene Copy Number Variation,' *Nature Genetics* 39, 10: 1256–1260, 2007.
8   Tishkoff, S.A., et al., 'Convergent Adaptation of Human Lactase Persistence in Africa and Europe,' *Nature Genetics* 39, 1: 31–40, 2007.
9   Owens, B., 'Genomics: The Single Life Sequencing DNA from Individual Cells is Changing the Way that Researchers Think of Human as a Whole,' *Nature* 491, 7422: 27–29, 1 November 2012.
10 Bergson, H., *Creative Evolution*, tr. A. Mitchell, Macmillan 1920.
11 Rossslenbroich B. *On the Origin of Autonomy: A New Look at the Major Transitions in Evolution,* Springer 2014.
12 Kandel, Eric R., *In Search of Memory: The Emergence of a New Science of Mind,* W.W. Norton & Co, New York 2007.
13 Amiel, J.J., Tingley, R. & Shine, R., 'Smart Moves: Effects of Relative Brain Size on Establishment Success of Invasive Amphibians and Reptiles,' *PLoS ONE* 6 (4) e 18277, 2011.
14 Bastuji, H., et al., 'Filtering the Reality: Functional Dissociation of Lateral and Medial Pain Systems during Sleep in Humans,' *Human Brain Mapping* 33 (11): 2638–2649, 2012.
15 Caceres, M., et al., 'Elevated Gene Expression Levels Distinguish Human from Non-Human Primate Brains,' *PNAS* 100, 22: 1303–1305, 2003.

Chapter 10

1   Wittgenstein, L., *Philosophische Untersuchungen*, in *Schriften*, Vol. 1: 279–544, Suhrkamp, Frankfurt 1969.
2   Libet, B., 'Unconscious Cerebral Initiative and the Role of Conscious Will in Voluntary Action,' *Behavioral and Brain Sciences* 8: 529–566, 1985.
3   Mieras, M., '"Spontane" beslissing is voorbereid ['Spontaneous' decision is prepared], *Volkskrant* January 28, 2012.
4   Smith, K., 'Neuroscience versus Philosophy: Taking Aim at Free Will,' *Nature* 477: 23–25, 2011. Comments included.
5   Soon, C.S., et al., 'Unconscious Determinants of Free Decisions in the Human Brain,' *Nature Neuroscience* 11, 5: 543–545, 2008.
6   Vohs, K.D. & Schooler, J.W., 'The Value of Believing in Free Will: Encouraging a Belief in Determinism Increases Cheating,' *Psychological Science* 19 (1): 49–54, 2008.
7   Crick, F.H.C., *The Astonishing Hypothesis. The Scientific Search for the Soul,* Simon & Schuster, London 1994.

8 Nichols, S., 'Free Will versus the Programmed Brain,' *Scientific American* 157, 19 August 2008.
9 Heinrich, J., Heine, S.J. & Norenzayan, A., 'The Weirdest People in the World?' *Behavioral and Brain Sciences* 33, 2/3: 61–135, 2010.
10 Mazar, N., On, A. & Ariely, D., *(Dis)Honesty: A Combination of Internal and External Rewards*. Working paper, Sloan School of Management, MIT, Cambridge, Mass. 2005.
11 Mazar, N. & Ariely, D., 'Dishonesty in Everyday Life and Its Policy Implications,' *Journal of Public Policy and Marketing* 25 (1), 2006.
12 Stillman, T.F., Baumeister, R.F. & Vohs, K.D., 'Personal Philosophy and Personnel Achievement: Belief in Free Will Predicts Better Job Performance,' *Social Psychological and Personality Science* 1: 43–50. 2010.
13 Baumeister, R.F., Masicampo, E.J., & DeWall, C.N., 'Prosocial Benefits of Feeling Free: Disbelief in Free Will Increases Aggression and Reduces Helpfulness,' *Personality and Social Psychology Bulletin* 35: 260–268, 2009.
14 Hermann, C., et al., 'Analysis of a Choice-Reaction Task Yields a New Interpretation of Libet's Experiments,' *International Journal of Psychophysiology* 67: 151–157, 2008.
15 Crick, F.H.C., *The Astonishing Hypothesis: The Scientific Search for the Soul*, Simon & Schuster, London 1994.
16 Dennett, Daniel, *Freedom Evolves*, Penguin, London 2003.
17 Tallis, R., *Aping Mankind. Neuromania, Darwinitis and the Misrepresentation of Mankind*, Acumen, Durham 2011.
18 Libet, B., 'Can Conscious Experience Affect Brain Activity?' *Journal of Consciousness Studies* 10, 12: 24–28, 2003.
19 Libet, B., 'Reflections on the Interaction of the Mind and Brain,' *Progress in Neurobiology* 78: 322–326, 2006.
20 *Volkskrant* November 17, 2012.
21 Milgram, S., 'Some Conditions of Obedience and Disobedience to Authority,' *Human Relations* 18, 57: 57–76, 1965.
22 Haney, C., Banks, W.C. & Zimbardo, P.G., 'Interpersonal Dynamics in a Simulated Prison,' *International Journal of Criminology and Penology* 1: 69–97, 1973.
23 Can be viewed on www.youtube.com as 'The Marshmallow Experiment'.
24 Mischel, W. & Ayduk, O., 'Willpower in a Cognitive-Affecting Processing System: The Dynamics of Delay of Gratification,' in Baumeister, R.F. & Vohs, K.D. (ed.), *Handbook of Self-Regulation: Research, Theory and Applications*, Guildford, New York 2004.
25 Tangney, J.P., Baumeister, R.F. & Boone, A., 'High Control Predicts Good Adjustment, Less Pathology, Better Grades and Interpersonal Success,' *Journal of Personality* 72: 271–324, 2004.
26 Tiemeijer, W.L., *Hoe Mensen Keuzes Maken* [How people make choices], Amsterdam University Press 2011.
27 Moffit, T.E., et al., 'A Gradient of Childhood Self-Control Predicts Health, Wealth and Public Safety,' *PNAS* 108: 2693–2698, 2011.
28 De Bruin, E. & Hijmans, A., *Terri Moffit and Avsalom Caspi: Zelfbeheersing en de kunst van het nee zeggen* [Self-control and the art of saying no], Volkskrant Boekenfonds 2012.

29 Dennett, D.C., *Elbow Room: The Varieties of Free Will Worth Wanting*, MIT Press, Cambridge, Mass. 1984.
30 Dijksterhuis, A., *Het Slimme Onbewuste: Denken met Gevoel* [The smart unconscious: thinking with feeling], Bert Bakker, Amsterdam 2008.
31 Starckx, S., 'Het onbewuste is modieus geworden' [The unconscious has become fashionable], *Psyche en Brein* 1, 2010.
32 Baumeister, R.F., et al., 'Ego Depletion: Is the Active Self a Limited Resource?' *Journal of Personality and Social Psychology* 74, 5: 1252–1265, 1998.
33 Tiemeijer, W. L., *Hoe Mensen Keuzes Maken* [How people make choices], Amsterdam University Press 2011.
34 Baumeister, R.F., et al., 'Self-Regulation and the Executive Function: the Self as a Controlling Agent,' in Kruglanski & Higgins (Ed.), *Social Psychology: Handbook of Basic Principles*, 2nd ed., Guildford, New York 2007.
35 Baumeister, R. & Tierney, J. *Willpower. Rediscovering the Greatest Human Strength*, Penguin, London 2011.
36 Lurquin J.H., et al., 'No Evidence of the Ego Depletion Effect across Task Characteristics and Individual Differences: a Pre-Registered Study,' *PLoS ONE* February 10, 2016.
37 Danziger, S., et al. 'Extraneous Factors in Judicial Decisions,' *PNAS* 108: 6889–6892, 2011.
38 Kahneman, D., *Thinking, Fast and Slow*, Farrar, Straus & Giroux, New York 2011.
39 Englich, B., et al., 'Playing Dice with Criminal Sentences: The Influence of Irrelevant Anchors on Experts – Judicial Decision Making,' *Personality and Social Psychology* 81: 657–669, 2001.
40 Lamme, V., *De Vrije Wil Bestaat Niet: Over wie er Echt Baas is in het Brein* [Free will does not exist: about who is the true boss in the brain], Bert Bakker, Amsterdam 2010.
41 Ibid.
42 Schiller, F., *On the Aesthetic Education of Man*, tr. Reginald Snell, Dover Publications, New York 2004.
43 Lamme, V., *De Vrije Wil Bestaat Niet: Over wie er Echt Baas is in het Brein*, Bert Bakker, Amsterdam 2010.

Chapter 11

1 Dawkins, R., *The Blind Watchmaker*, W.W. Norton & Co., New York 1986.
2 Sheldrake, R., *The Science Delusion. Freeing the Spirit of Enquiry*, Coronet, London 2012.
3 Webb, P., 'The Measurement of Energy Exchange in Man: An Analysis,' *American Journal of Clinical Nutrition* 33: 1299–1310, 1980; and 'The Measurement of Energy Expenditure,' *Journal of Nutrition* 121: 1897–1901, 1991.
4 Sheldrake, R., *The Science Delusion. Freeing the Spirit of Enquiry*, Coronet, London 2012.
5 Dawkins, R., *The Selfish Gene*, Oxford University Press 1976.
6 Sheldrake, R., *The Science Delusion. Freeing the Spirit of Enquiry*, Coronet, London 2012.

7 Fortey, R., *Life: A Natural History of the First Four Billion Years of Life on Earth*, Random House, New York 1997.
8 Freeland, S.J. & Hurst, L.D., 'Evolution Encoded,' *Scientific American* 290, April 2004.
9 Fuchs, T., *Das Gehirn – ein Beziehungsorgan: Eine phänomenologisch-ökologische Konzeption*, Kohlhammer, Stuttgart 2010.
10 Dennett, D.C., *Consciousness Explained*, Black Bay Books 1992.
11 Ibid.
12 Nagel, T., *Mind and Cosmos: Why the Materialist Neo-Darwinian Conception of Nature is Almost Certainly False*, Oxford University Press 2012.
13 Leuchter, A.F., et al., 'Changes in Brain Function of Depressed Subjects During Treatment with a Placebo,' *American Journal of Psychiatry* 159: 122–129, 2002; and Mayberg, H.S., et al., 'The Functional Neuroanatomy of the Placebo Effect,' *American Journal of Psychiatry* 159: 728–736, 2002.
14 Aldenhoff, J., 'Überlegungen zur Psychobiologie der Depression,' *Nervenartzt* 68: 379–389, 1997.
15 Heijden, M. van der, 'Dick Swaab: Een goede tegenwerping heb ik nog niet gehoord' [Dick Swaab: I have yet to hear a good objection], *NRC Leesclub* November 5, 2011.
16 Kahn, R., *Perspectief,* Bunge, Utrecht 1994.
17 Cabanis, P.J.G., *Rapports du physique et du moral de l'homme*, part 1, 1802, p. 151.
18 Swaminathan, N., 'Slumber Reruns: As We Sleep, Our Brains Rehash the Day. And Six Times Faster than the Normal Speed,' *Scientific American* November 2007.
19 Dworak, M., et al., 'Sleep and Brain Energy Levels: ATP Changes during Sleep,' *Journal of Neuroscience* 30, 26: 9007–9016, 2010.
20 Wong-Riley, M., 'What Is the Meaning of the ATP Surge during Sleep?' *Sleep* 34, 7: 833–834, 2011.
21 Khamsi, R., 'Energetic Cells May Have Boosted the Brain. Did Rapid Mutation of Cell Powerhouse Guide Our Neural Evolution?' *News@Nature.com* November 30, 2004.
22 Sorger, B., et al., 'A Real-Time fMRI-Based Spelling Device Immediately Enabling Robust Motor-Independent Communication,' *Current Biology* 22, 14: 1333–1338, 2012.

Chapter 12

1 Templeton, J.J., et al., 'In the Eye of the Beholder: Visual Mate Choice Lateralization in a Polymorphic Songbird,' *Biology Letters* 8 (6): 924–927, 2012.
2 Fisher, H., et al., 'Romantic Love: an fMRI Study of a Neural Mechanism for Mate Choice,' *Journ. Comp. Neurol.* 493, 1: 58–62, 2005.
3 Thiebaut de Schotten, M., et al., 'A Lateralized Brain Network for Visuospatial Attention,' *Nature Neuroscience* 14:1245–1246, 2011.
4 Ibid.
5 Schore, A.N., 'Back to Basics: Attachment, Affect Regulation and the Developing Right Brain: Linking Developmental Neuroscience to Pediatrics,' *Pediatrics in Review* 26 (6): 204–217, 2005.

6 Schore, A.N., 'A Neuropsychoanalytic Viewpoint: A Commentary on a Paper by Steven H. Knoblauch,' *Psychoanalytic Dialogues* 15 (6): 829–854, 2005.
7 Doidge, N., *The Brain that Changes Itself. Stories of Personal Triumph from the Frontiers of Brain Science*, Penguin, London 2008.
8 Ingalhalikar, M., et al., 'Sex Differences in the Structural Connectome of the Human Brain,' *PNAS* (online before print), December 2, 2013.

Chapter 13

1 Hoornik, Ed., *Het Menselijke Bestaan* [Human existence], Daamen's, The Hague 1952.
2 Bolte Taylor, J., *My Stroke of Insight. A Brain Scientist's Personal Journey*, Viking Penguin, New York 2006.
3 McGilchrist, I., *The Master and His Emissary. The Divided Brain and the Making of the Western World*, Yale University Press 2009.
4 Ibid.
5 Shamay, S. G., et al., 'The Neuroanatomical Basis of Understanding of Sarcasm and its Relationships to Social Cognition,' *Neuropsychology* 19, 3: 288–300, 2005.
6 Kounios, J., et al., 'The Prepared Mind: Neural Activity Prior to Problem Presentation Predicts Solution by Sudden Insights,' *Psychological Science* 17: 882–980, 2006; and Beeman M.-J., et al., 'Neural Activity Observed in People Solving Verbal Problems with Insight,' *Public Library of Science-Biology* 2: 500–510, 2004.
7 McGilchrist, I., *The Master and His Emissary. The Divided Brain and the Making of the Western World*, Yale University Press 2009.
8 Ibid.
9 ACCORD trial 2008 and 2010, see Ledford, H., *Cholesterol Limits Lose Their Lustre*, in: Nature 494: 410–411, February 2013.
10 Doidge, N., *The Brain That Changes Itself. Stories of Personal Triumph from the Frontiers of Brain Science*, Penguin, London 2007.
11 Donald, M., 'The Central Role of Culture in Cognitive Evolution: A Reflection on the Myth of the "Isolated Mind",' in Nucci, L. (Ed.), *Culture, Thought and Development*, Lawrence Erlbaum, New Jersey 2000.
12 Nisbett, R. E., et al., 'Culture and Systems of Thought: Holistic versus Analytic Cognition,' *Psychological Review* 291–310, 2001.
13 Cabeza, R., 'Hemispheric Asymmetry Reduction in Older Adults: the HAROLD Model,' *Psychology and Aging* 17 (1): 85–100, 2002.

Chapter 14

1 Groot A. D. de, *Thought and Choice in Chess*, Amsterdam University Press 1965.
2 Dijksterhuis, A., *Het slimme onbewuste. Denken met gevoel* [The smart unconscious: thinking with feeling], Bert Bakker, Amsterdam 2008.
3 Kahneman D. *Thinking, Fast and Slow*. Penguin, London 2012.
4 Bertrand, M., 'Implicit Discrimination,' *American Economic Review* 95, 2: 94–98, 2005.
5 Baumeister, R., *Willpower, Rediscovering the Greatest Human Strength*, Penguin Putnam, New York 2011.

NOTES

6   Petersen, S.E., et al., 'Positron Emission Tomography Studies of the Cortical Activity of Single-Word Processing,' *Nature* 331, 585–589, 1988.
7   Lurquin, J.H., et al., 'No Evidence of the Ego-Depletion Effect across Task Characteristics and Individual Differences: A Pre-Registered Study,' *PLoS ONE* February 10, 2016.
8   Schwartz, J., et al., 'Systematic Changes in Cerebral Glucose Metabolic Rate after Successful Behaviour Modification Treatment of Obsessive Compulsive Disorder,' *Arch. Gen. Psychiatry* 53: 109–113, 1996; and Paquette, V., et al., 'Change the Mind and You Change the Brain: Effects of Cognitive-Behavioural Therapy on the Neural Correlates of Spider Phobia,' *Neuroimage* 18: 401–409, 2003; and Beauregard, M., et al., 'Mind Really Does Matter: Evidence from Neuroimaging Studies of Emotional Self-Regulation, Psychotherapy and Placebo Effect,' *Progress in Neurobiology* 81: 482–761, 2007.

Chapter 15

1   James, W., *The Principles of Psychology*, Macmillan London 1890.
2   Cyrianoski, D., 'Neuroscience: The Mind Reader,' *Nature* 486: 178–180, 2012.
3   Owen, A.M., et al., 'Residual Auditory Function in Persistent Vegetative State: A Combined PET and fMRI Study,' *Neuropsychological Rehabilitation* 15: 290–306, 2005.
4   Monti, M.M., 'Willful Modulation of Brain Activity on Disorders of Consciousness,' *New England Journal of Medicine* 362: 579–589, 2010.
5   Nahm, M. & Greyson, B., 'Terminal Lucidity in Patients with Chronic Schizophrenia and Dementia: A Survey of the Literature,' *Journal of Nervous and Mental Diseases* 197, 12: 943–944, 2009; and Fenwick, P., 'End-of-Life Experiences: Reaching out for Compassion, Communication and Connection-Meaning of Deathbed Visions and Coincidences,' *American Journal of Hospice and Palliative Medicine* 28, 1: 7–15, 2011; and Nahm, M., et al., 'Terminal Lucidity: A Review and a Case Collection,' *Archives of Gerontology and Geriatrics* 55,1: 138–142, 2012.
6   Kelly, E.F., *Irreducible Mind. Toward a Psychology for the 21st Century*, Roman & Littlefield, Plymouth 2007.
7   Nahm, M. & Greyson, B., 'Terminal Lucidity in Patients with Chronic Schizophrenia and Dementia: A Survey of the Literature,' *Journal of Nervous and Mental Diseases* 197, 12: 943–944, 2009.
8   Osis, K. & Haraldson, E., *At the Hour of Death*, Hastings House, Norwalk, Conn. 1997.
9   Turetskaia, B.E. & Romanenko, A.A., 'Agonal Remission in Terminal Stages of Schizophrenia,' *Journal of Neural Pathology and Psychiatry* 75: 559–562, 1975.
10 Grosso, M., *Experiencing the Next World Now*, Paraview, New York 2004.
11 Lewin, R., 'Is Your Brain Really Necessary?' *Science* 210: 1232–1234, 1980; and Lorber, J., 'Is Your Brain Really Necessary?' in Voth, D. (Ed.), *Hydrocephalus in frühen Kindesalter: Fortschritte der Grundlagenforschung, Diagnostik und Therapie*, Enke, Stuttgart 1983, pp. 2–14.
12 'Minerva' *BMJ* 327, p. 998, 2003 (by Yuen and Green).
13 Kossof, E.H., et al., 'Hemispherectomy for Intractable Unihemispheric Epilepsy Etiology vs Outcome,' *Neurology* 62, 7: 887–890. 2003.

14 Choi, C., 'Strange but True: When Half a Brain Is Better than a Whole One,' *Scientific American* May 24, 2007.
15 This story is hard to trace online, but you can read other similar accounts at www.sturgeweber.org.uk.
16 Borgstein, J. & Grootendorst, C., 'Half a Brain,' *Lancet* 359, 9305: 473, 2002.
17 Shewmon, D.A., 'Chronic "Brain Death": Meta-Analysis and Conceptual Consequences,' *Neurology* 51: 1538–1545, 1998.
18 Sperling, D., *Management of Post-Mortem Pregnancy: Legal and Philosophical Aspects,* Ashgate, Aldershot 2006.
19 Kerkhoffs, J., *Droomvlucht in coma* [Dream flight in coma], Melick 1994. Autobiographical story of a man declared brain dead who, because the family refused to permit donation of his organs, was disconnected from the oxygen machine and woke up. Disconnection (apnea test) is part of current brain death protocol. Esmee Feenstra was in 2005 declared brain death. Just before the apnea test her sister saw a tear coming out of her eyes so the procedure was stopped. Esmee has completed an academic study since then. (http://orgaandonorjaofnee.nl/links/)
20 There are two cases of brain dead people waking up in the USA: Zack Dunlap, brain dead in accordance with fully executed protocol, came to just before his organs were to be harvested. (youtube.com/watch?v=u6CXKBYiNKs and Kupferschmidt, R. & White, H., 'Woman Diagnosed as Brain Dead Walks and Talks after Awakening,' *LifeSiteNews,* February 15, 2008). There were also two cases in England, the teenager Steven Thorpe: Furness., H., 'Miracle Recovery of Teen Declared Brain Dead by Four Doctors,' *The Telegraph* April 25, 2012; and Colleen Burns: Saul, H., 'Dead Woman Wakes up as her Organs are About to be Harvested,' *The Independent* July 9, 2013.
21 Verwer, R.W.H., et al., 'Cells in Post-Mortem Human Brain Tissue Slices Remain Alive for Several Weeks in Culture,' *Faseb Journal* 16: 45–60, 2002.
22 Evers, M., 'Back from the Dead: Resuscitation Expert Says End is Reversible,' *Der Spiegel* (online) July 29, 2013; www.spiegel.de/international/world/doctor-sam-parnia-believes resurrection-is-a-medical-possibility-a-913075.html
23 Parnia, S. & Young, J., *Erasing Death: The Science that is Rewriting the Boundaries between Life and Death,* Harper Collins, New York 2013.
24 Sacks, O., *Awakenings,* Picador, Pan Books, London 1982.
25 Truog, R.D. & Miller, F.G., 'The Dead Donor Rule and Organ Transplantation,' *New England Journal of Medicine* 359, 7: 674–675, 2008.
26 President' Council on Bioethics, *Controversies in the Determination of Death,* Washington DC 2008.
27 Thonnard, M., et al., 'Characteristics of Near-Death Experience Memories as Compared to Real and Imagined Event Memories,' *PLoS ONE* 8 (3): e57620. doi.10.1371/journal.pone.0057620 (2013).
28 Brumfield, B., '"Afterlife" Feels "Even More Real than Real," Researcher says,' *CNN Labs* April 10, 2013 (html://edition.cnn.com/2013/04/09/health/belgium-near-death-experiences/index.html?hpt=hp_c5).
29 Lommel, P. van et al., 'Near-Death Experience in Survivors of Cardiac Arrest: A Prospective Study in the Netherlands,' *Lancet* 358, 9298: 2039–2045, 2001.
30 Lommel, P. van, *Consciousness Beyond Life: The Science of the Near-Death Experience,* HarperCollins. 2011.

31. Sabom, M., *Light and Death: One Doctor's Fascinating Account of Near-Death Experiences,* Zondervan, Mich. 1998.
32. Alexander, E., *Proof of Heaven. A Neurosurgeon's Journey into the Afterlife,* Simon & Schuster, New York 2012.
33. Nuzo, R., 'Babies' Brains May be Tuned to Language Before Birth,' *Nature* doi: 10.1038/Nature 2013.12489.
34. Prechtl, H.F.R., 'Ultrasound Studies of Human Fetal Behaviour,' *Early Human Development* 12 (2): 91–98, 1985.
35. Hill, L.M., Platt, L.D. & Manning, F.A., 'Immediate Effect of Amniocentesis of Fetal Breathing and Gross Body Movement,' *American Journal of Obstetrics and Gynecology* 35: 689–690, 1979.
36. Chamberlain, D., *Windows to the Womb. Revealing the Conscious Baby from Conception to Birth,* North Atlantic Books, Berkeley 2013.
37. Arabin, B., et al., 'The Onset of Inter-Human Contacts: Longitudinal Ultrasound Observations in Early Twin Pregnancies,' *Ultrasound Obstetrics & Gynecology* 8 (3): 166–173, 1996.
38. Piontelli, A., 'A Study on Twins before and after Birth,' *International Review of Psycho-Analysis* 16: 413–426, 1989.
39. Liley, A.W., 'The Foetus as a Personality,' *Australia and New Zealand Journal of Psychiatry* 6 (2): 99–105, 1972.

Chapter 16

1. Can be viewed on www.youtube.com as 'Three ingredients for murder: James Fallon'.
2. Can be viewed on www.youtube.com as 'The rubber hand illusion'.
3. Botvinick, M. & Cohen, J., 'Rubber Hands Feel "Touch" that Eyes See,' *Nature* 391: 756, 1998.
4. Shokur, S., 'Expending the Primate Body Schema in Sensimotor Cortex by Virtual Touches of an Avatar,' *PNAS* online, August 26, 2013.
5. Hume, D., 1, 4, 6 (1739–1740), *Treatise of Human Nature,* Penguin, London 1996.
6. Heijden, M. van der, 'Ergens tussen zojuist en straks' [Somewhere between a moment ago and soon], *NRC Weekend* October 26–27, 2013.
7. Augustine, St, *City of God,* Ch. 26. From *Library of the Nicene and Post-Nicene Fathers of the Church,* www.ccel.org/ccel/schaff/npnf102.iv.XI.26.html.
8. Dennett, D., *Consciousness Explained,* Little, Brown and Company 1991.
9. Noë, A., *Out of Our Heads. Why You Are Not Your Brain, and Other Lessons from the Biology of Consciousness,* Hill & Wang, New York 2009.
10. Raichle, M.E., 'Inaugural Article: A Default Mode of Brain Function,' *PNAS* 98: 676–682, 2001.
11. Mason, F.M., et al., 'Wandering Mind: the Default Network and Stimulus Independent Thought,' *Science* 19, 315 (5810): 393–395, 2007.
12. Greicius, M.D., et al., 'Default Mode Network Activity Distinguishes Alzheimer's Disease from Healthy Aging: Evidence from Functional MRI,' *PNAS* 101, 13:4637–4642, 2004.
13. Sheline, Y.I., Raichle, M.E., et al., 'The Default Mode Network and Self-Referential Processes in Depression,' *PNAS* 106, 6: 1942–1947. 2009.

14 Hurk, P.A. van der, et al., 'An Investigation of the Role of Attention in Mindfulness-Based Cognitive Therapy for Recurrently Depressed Patients,' *Journal of Experimental Psychopathology* 3, 1: 103–120, 2012.
15 Hopkin, M., 'Meditating Monks Focus the Mind. Buddhists Show Clarity of Attention in Optical Illusion Tasks,' *News@Nature.com* 8 June, 2005.
16 Bellow, S., *Humboldt's Gift,* Penguin, London 1973.
17 Oliner, S. & Oliner, P., *Altruistic Personality,* Simon & Schuster, New York 1992.
18 Ledoux, J., *The Synaptic Self: How Our Brains Become Who We Are,* Viking, London 2002.
19 Bos, A., *Hoe de Stof de Geest Kreeg* [How matter got spirit], Christofoor, Zeist 2008.
20 Anonymous, Neuroscience: My Life with Parkinson's, *Nature* 503: 29–30, 2013.

Chapter 17

1 Goethe, J. W., *Faust II,* Act 1, tr. David Constantine, Penguin 2009.
2 Dennett, D.C., *Darwin's Dangerous Idea: Evolution and the Meanings of Life,* Simon & Schuster, New York 1995.
3 Dawkins, R., 'Let's All Stop Beating Basil's Car' www.edge.org/response-detail/11416
4 Watson, J.B., *Behaviorism,* University of Chicago Press (revised ed.) 1930.
5 Poll, W. van de, 'Onze neuronen zijn ons te snel af' [Our neurons are too quick for us], *Trouw* February 5, 2011.
6 Waal, F. de, *Good Natured: The Origins of Right and Wrong in Humans and Other Animals,* Harvard University Press, Cambridge, Mass. 1996.
7 Quoted in Waal, F. de, *Primates and Philosophers: How Morality Evolved,* Princeton University Press 2006.
8 Quoted in Dekker, C., *Het kleine is groots,* [Small is great], inaugural lecture, Delft November 17, 2000.
9 Merleau-Ponty, M., *Phénoménologie de la perception,* Gallimard, Paris 1945.
10 Denys D. & Meynen G. *Handboek Psychiatrie en Filosofie.* De Tijdstroom, Utrecht. 2011.
11 Maassen, H., 'Ziekte moet verteld worden' [Illness has to be related in words], *Medisch Contact* 68, 13: 652–654, 2013.
12 Goethe, J.W. von, *The Metamorphosis of Plants,* MIT Press 2009.
13 Dennett, D.C., *Freedom Evolves,* Penguin Books, London 2003.
14 Gazzaniga, M.S., *The Social Brain: Discovering the Networks of the Mind,* Basic Books, New York 1987.
15 Sedlmeier, P., et al., 'The Psychological Effects of Meditation: A Meta-Analysis,' *Psychological Bulletin* 138 (6): 1139–1171, November 2012.

Index

Aarsman, Hans 21
Alexander, Eben 197
altruism 8, 112, 222
Alzheimer's 185
amniocentesis 200
amputations 67
Andel, Hendrikje van 160
ANP (atrium natriuretic peptide) 98
antibiotics 100
antidepressants 144
apes 111, 155
aping 31
Arabin, Birgit 201
area striata 41
Ariely, Dan 118
Aristotle 141
ATP (adenosine triphosphate) 147
attention 78
auditory cortex 72
Augustine, St 206
autism 34, 38, 154
automatons 8, 133, 175, 221, 223

babies, new-born 18, 30
Bach-y-Rita, George 24
Bach-y-Rita, Paul 26, 73f
Bach-y-Rita, Pedro 24f, 69, 163
Bartels, Andreas 57
Bastiaansen, Jojanneke 38
Baumeister, Roy 127, 177
Beauregard, Mario 57
behaviour,
—, behaviourism 221
—, inherited 102
—, unethical 119
—, utilisation 133

Bellow, Saul 212
Bennett, Craig 53f
Bennett, Max 74
benzodiazepides 99
Bergson, Henri 12, 81, 104, 197
Beuys, Joseph 53
binding problem 65
Bingen, Hildegard von 89
blood 60, 162, 168
body schema 93
body-spirit dualism 76
Bois-Reymond, Emil du 137, 165
Bolk, Louis 228
Bolte Taylor, Jill 162
brain,
— death 191, 193
— development 15
— haemorrhage 162
—, left 163f, 166f, 169f
—, right 163f, 168–70
Brentano, Franz 226
Broca's area 18f, 33
Brücke, Ernst von 137, 165

Cabanis, Pierre 145
callotomy 189
Caspi, Avsalom 125
caudate nucleus 57
causal chains 121
cerebrospinal fluid 187
Chalmers, David 145
chicken 152
childhood 172
cholesterol 168
Chomsky, Noam 33
chronic pain syndrome 67, 69

Churchill, Winston 132
Churchland, Patricia 65
cingulate cortex (gyrus) 62, 109, 210
—, anterior 57
coma 183, 188, 197
connections, plasticity of 15
consciousness 9, 48, 58, 65, 79, 81f, 127, 182, 199
—. minimal 194
—, self-reflecting 7, 178
corpus callosum 151
cortex 110
Crick, Francis 13, 117, 140
Cruyff, Johan 151, 161
cryonics 87
culture 111f, 160, 170
cytokines 98

Dalrymple, Theodore 12
Damasio, Antonio 76, 89, 94
Darwinism 101, 232
Dawkins, Richard 137, 139, 220f
daydreaming 211
de-individuation 124, 134
dementia 185
Dennett, Daniel 76, 84, 121, 123, 126, 142, 208, 214, 219, 221, 224, 229
Denys, Damiaan 178
depression 92, 98, 179
Derrida, Jacques 208
Descartes, René 75, 138, 141, 167, 206

diabetes 168
Dijksterhuis, Ap 126, 176, 215
dissociative identity syndrome 40
DMN (default mode network) 210
DNA 13, 23, 102, 140, 228, 233
dog 111
Doidge, Norman 25, 159
dopamine 92, 166
Dunedin 125
dyslexia 154

Eeden, Frederik van 50, 53
EEG (electro-encephalogram) 44, 50, 77, 164, 191
ego depletion 127, 177
Eldredge, Niles 228
embodied cognition 95, 97
emotions 89–91, 165, 234
empathy 34f, 112, 176
encephalisation 108
environment 104
epiphenomenalism 143
Ericsson, K. Anders 17, 26
ethics 222
Euclid 167
evolution 86, 101f, 105, 113
experience 226
eyes 152

facial
— expression 92
— recognition 20, 165
Fallon, James 202, 213
feelings 89, 144, 146
Feinstein, Bertram 63
ferrets 73
Fisher, Helen 57
fly 7
fMRI (functional Magnetic Resonance Imaging) 31, 50, 52-6, 90, 117, 164, 182
—, BOLD (blood-oxygen-level dependent) 59
foetus 201
Fortey, Richard 140
free will 8, 111, 114, 118, 215, 217, 229
frog experiments 137
frontal lobe 110f, 133
Fuchs, Thomas 48, 81, 142, 226

GABA (gamma-amino-butyric-acid) 78
Galileo 141
Gallese, Vittorio 226
Gazzaniga, Michael 233
gender 159
genes 102f
Gestalt 165
Goethe, Johann Wolfgang von 219, 228
Gould, Stephen 228, 230
—, Gouldian finches 153
Grayling, Anthony 86
Greenfield, Susan 145
Grossmann, Lawrence 148
Grossman, Vasily 101
growth patterns 228
gyrus
— angularis 34, 93
— cinguli 62, 109, 210
— fusiformis 20
— marginalis 34
— temporalis inferior 20

habituation 107
Hacker, Peter 74
Haens, Geert d' 99
Haldane, J.B.S. 47
Hameroff, Stuart 224
HAROLD model 160, 172
Haynes, John-Dylan 116f
Head, Henry 67, 69
Hebb, Donald O. 17

Helmholtz, Hermann von 137f
hemisphere 151, 163, 166–69, 234
hemispherectomy 189
Hermann, Christoph 120
hippocampus 23, 210
HIV virus 103, 172
Hoffer, Eric 29
Hofstadter, Douglas 123
Homo
— erectus 148
— habilis 101
— heidelbergensis 155
— sapiens 104, 112, 148, 155
homunculus 74, 76, 206
Hoornik, Ed. 162
Hopf, J.M. 45
hormones 98
human and chimp 103
Human Brain Project 87
Hume, David 205, 212
Husserl, Edmund 226
Huxley, Thomas 143
hydrocephaly 187
hysteria 134

idiots savants 197
imitation 30
—, behaviour 134
immune system 97
inferior parietal lobule 155
inner dialogue 162
insula 35
intelligence 23
intelligent design 105, 143
invisible gorilla 79
inferior parietal lobe (IPL) 155

James, William 80, 89, 182
Jansen, Jan 36
Johnson, Mark 94

Kahneman, Daniel 129, 176
Kahn, René 92, 145

INDEX

Kandel, Eric 106
knowledge, implicit 165
Kousbroek, Rudy 232

Lacan, Jacques 208
Lakoff, George 94
Lamarck, Jean-Baptiste 102
Lamme, Victor 51f, 119, 127, 131f, 175, 206, 209, 221
Lashley, Karl Spencer 71
Laureys, Steven 183, 195
LeDoux, Joseph 23, 129, 214
Libet, Benjamin 62, 64, 66, 80, 115, 122, 233
Liley, Sir William 201
locked-in syndrome 182
lollypop 26, 73
Lommel, Pim van 196
Lorber, John 186
LSD 170

maladaptive plasticity 68
Maslach, Christina 124
materialism 140–42, 172, 209, 219, 223f, 231
Matrix, The 85
McGilchrist, Iain 153, 157, 164, 166, 169, 226, 233
McLuhan, Marshall 51
medial insula 57
meditation 212, 234
memory 211
meningitis 191, 197
Merleau-Ponty, Maurice 61, 81, 226, 228
metaphors 94
microtubuli 224
Mieras, Mark 115, 120
Milgram, Stanley 123
mind 9
mindfulness 212
mitochondria 148
Moffit, Terri 125
Mokkenstorm, Jan 179

monkeys 19, 30f, 33, 35, 111, 139, 205
Montagnier, Luc 173
morality 112
Mosley, Michael 26, 61, 67, 222
mother tongue 18
Mountcastle, Vernon 73
MRI (Magnetic Resonance Imaging) 116
Mulder, Theo 69
Müller, Johannes 137
Müller-Leyer illusion 21f, 118, 130, 167
Munnik, Martin de 222
Muotri, Allyson 104
myelin 16, 111, 188

Nagel, Thomas 54, 143, 222
Nazism 232
NDE (near-death experience) 195f, 199
Neanderthals 155
neglect 156
nerve cells 70
nerves, sensory 107
neurodeterminism 7f, 196, 223
neurons 17, 23, 104, 224
—, canonical 30
—, mirror 29–37, 92, 94, 111, 176, 213, 227
neuroplasticity 108
neurotransmitters 92
Nietzsche, Friedrich 29, 86, 166, 169
Nisbett, Richard E. 171
Noë, Alva 21, 73, 81, 209, 226
nucleus accumbens 92

Occam's razor 64
Oeveren, Efraim van 79
olfactory bulb 23
Oliner, Pearl 214
Oliner, Samuel 214
Orden, Guy van 70

O'Regan, Kevin 77
organisms 230
—, multicellular 105
—, unicellular 105, 141
Owen, Adrian 182
oxygen 193
oxytocin 92, 166

pain,
—, phantom 67–69
—, sympathetic 35
paraplegia 149
Parkinson's disease 63, 216
Parnia, Sam 192f, 196
parrots 36
Pascal, Blaise 174
pattern recognition 176
Pavlov, Ivan Petrovitch 221
peer pressure 124
Penrose, Sir Roger 224
phenomenology 226, 234
pheromones 166
pineal gland 75
Piontelli, Alessandra 201
plasticity 26, 28, 43, 160, 175
Poincaré, Henri 139
Polanyi, Michael 165
polyp (hydra) 106f
Popper, Karl 132
pre-tectum 135
projection 81
pseudo-science 133
psychoses 215
psychosomatics 97
punctuated equilibrium 229
Putnam, Hilary 84

Raichle, Marcus 210
Ramachandran, Vilayanur Subramania 30f, 33, 35, 68, 93, 155, 166
rats 71
readiness potential 120
reductionism 172
reification 220, 232
reincarnation 91

255

REM sleep 62, 200
retina 81, 134
Reynolds, Pam 197
Rigoni, David de 120
Rivers, William 67, 69
Rosslenbroich, Bernd 105, 229

Sacks, Oliver 21, 194, 228
Sautoy, Marcus de 116
savant syndrome 154
Schiller, Friedrich 134
schizophrenia 154
Schrödinger, Erwin 224
self 49, 60, 178f, 199, 204, 208, 212, 216, 225, 227, 230
self-control 111, 122, 124, 218
selfish gene 113
sensitisation 107
serotonin 92, 99
Sheldrake, Rupert 77, 139
Shewmon, Alan 191
Singer, Isaac Bashevis 115
Soltmann, Otto 72
Soon, Chun Siong 117
soul 90, 129, 233
speaking 18
Spinoza, Baruch 220
spirit 179, 233

Stanford marshmallow test 124
stress 98
stroke 182
Sturge-Weber syndrome 189
superstring theory 232
Swaab, Dick 6, 83, 85f, 119, 145, 149, 193
Sylvian fissure 155
synapses 103
system 1 129–31, 171, 176, 215, 218
system 2 129f, 171, 176, 184, 218

tacit knowledge 165
Tallis, Raymond 57, 122
tectum 135
terminal lucidity 184
theory of mind 209
thermodynamics 138
top-down-effect 44
tranquillisers 78, 144
transposons 23, 102, 104
twins 200

understanding 94

Venter, Craig 140
VEP, see Visually Evoked Potentials
veto power 204

visual cortex 44, 72, 108
visual illusions 74
Visually Evoked Potentials 41, 77
visual stimulus 74
vitalism 138, 140, 230
Voltaire 119
voxel (volume pixel) 51f
Vroon, Piet 233

Waal, Frans de 112, 222, 230
Waldvogel, Bruno 40
Wallot, Sebastian 70
Watson, James 13, 140
Watson, John Broadus 221
Webb, Paul 139
WEIRD categorisation 118
Wernicke's area 19, 33
Whitehead, Alfred North 81
Wittgenstein, Ludwig 115, 203
Wong-Riley, Margaret 147
Wunengzi 137

Yakovlev torque 154

Zeki, Semir 57
Zwagerman, John 179

Floris Books

For news on all our **latest books**, and to receive **exclusive discounts**, **join** our mailing list at:

florisbooks.co.uk

Plus subscribers get a FREE book with every online order!

We will never pass your details to anyone else.